BLAISE PASCAL

☞ Books in the RENAISSANCE LIVES series explore and illustrate the life histories and achievements of significant artists, intellectuals and scientists in the early modern world. They delve into literature, philosophy, the history of art, science and natural history and cover narratives of exploration, statecraft and technology.

Series Editor: François Quiviger

Already published

Blaise Pascal: Miracles and Reason *Mary Ann Caws*

Caravaggio and the Creation of Modernity *Troy Thomas*

Hieronymus Bosch: Visions and Nightmares *Nils Büttner*

Michelangelo and the Viewer in His Time *Bernardine Barnes*

BLAISE PASCAL

Miracles and Reason

MARY ANN CAWS

REAKTION BOOKS

To Dr Paige Ambroziak, for her inestimable assistance in the research, writing and translations for this book, in particular the scientific material but also much else, my warmest gratitude. Her knowledge and perseverance have been essential during the lengthy period of preparation for this homage to Pascal.

Published by Reaktion Books Ltd
Unit 32, Waterside
44–48 Wharf Road
London N1 7UX, UK
www.reaktionbooks.co.uk

First published 2017

Printed and bound in China by 1010 Printing International Ltd

A catalogue record for this book is available from the British Library

ISBN 978 1 78023 721 3

COVER: François Quesuel the Younger, *Blaise Pascal*, *c.* 1690, oil on canvas. Photo © akg-images/Fototeca Gilardi

CONTENTS

PREFACE

Living with Blaise

Tom Conley

IT IS TEMPTING to believe that Mary Ann Caws reads Pascal through 'Madeleine à la veilleuse', René Char's short poem of 1934 honouring Georges de La Tour's painting of the same name (illus. 1). Ineffably, through Char's words and the matte beauty of the painting where Mary Magdalene, as if the penitent were sculpted in wood, the bright flame of a burning wick in a glass of oil, reflects on the objects on a table that could be the components of a penitent still-life – a coiled scourge, two sealed books, a wooden cross – Caws might be thinking of the *Pensées*. Unlike legions of critics whose boundless admiration for Pascal sifts through the networks of erudition, Caws gets at him from the inside. Perhaps or perhaps not like the Madeleine who countenances time and death in candle-light, whose left hand buttresses her chin and whose right rests on the cold surface of the death's head coddled in her lap, Caws avows that she has spent a part of her life musing over Pascal. She quotes Lucien Goldmann: 'I am one of those whom Pascal upsets but doesn't convert. Pascal, the greatest of all, yesterday and today', and has a sudden memory-flash, a fragment that erupts from the past: on 6 November 1956,

1 Georges de La Tour, *Magdalene with the Smoking Flame*, c. 1640, oil on canvas.

at the age of 23, she recalls being so attracted to the wars of the Jesuits and Jansenists that when beginning her studies at the Institut catholique de Paris she imagined herself in the setting of *Provinciales*. Long after, she adds, when she encountered Roland Barthes' *Sade Loyola Fourier* (1976), her interest again caught fire.

This book makes manifest an unspoken passion anyone of French leaning will share with the *Pensées*. When, in his discussions about Port-Royal and the *Essais*, the literary botanist Sainte-Beuve admitted that 'il y a du Montaigne dans chacun de nous' (there is some Montaigne in all of us), both he and his audience were in most likelihood also thinking of Pascal. In unbounded admiration the visionary surgeon, leftist, art historian and cinephile Élie Faure counted Pascal, along with Shakespeare and Cervantes, among the essayist's first-born children. Early in his career the Marxian philosopher Henri Lefebvre discovered in Pascal the foundations for lifelong reflections on space, everyday life and the rhythms of bodily being.[1] Today Caws tells us that in the age of the digit, the mobile phone, the iPod and iPad, Instagram and Twitter, Pascal's styles of writing and thinking remain a supple and ever-current measure of an ethics of truth and action.

To obtain a sense of how the form *Blaise Pascal: Miracles and Reason* configures the space of its reflections it suffices to recall the way that Maurice Blanchot, another fervent reader of Pascal, organized *L'Espace littéraire*, an assemblage of essays on Mallarmé, Rilke, Kafka and Hölderlin. In a foreword Blanchot speculated that however fragmentary or scattered its history or aspect, a work of substance has a force of attraction drawing its readers towards a mobile centre.[2] One of

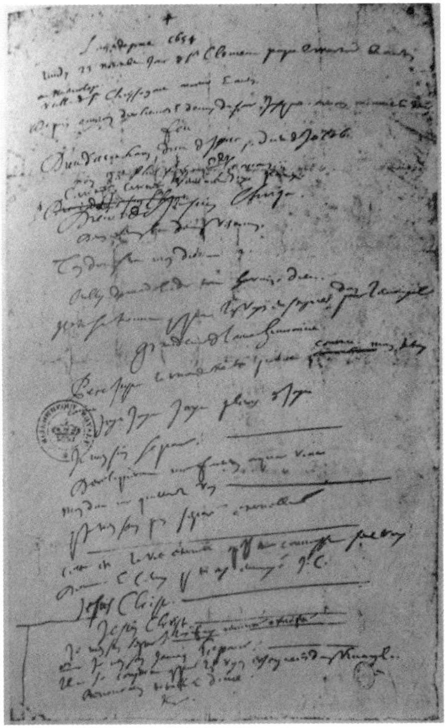

the moving axes or vanishing points of *Blaise Pascal* would be found in Chapter Five, where Caws situates and cites in full Pascal's famous Memorial of 23 November 1654. 'My first engagement,' she underscores, 'so very many years ago, with this very odd genius began with my reading of this text.'

In its very form the Memorial, the name the Oratorian Pierre Guerrier assigned to the piece, is an enduring enigma. In 1732, sixty years after Pascal's passing, Guerrier reported that in the days immediately following the polymath's death a domestic had noticed something bulging in the deceased's waistcoat. He was said to have cut the thread sealing the

2 Pascal's Memorial.

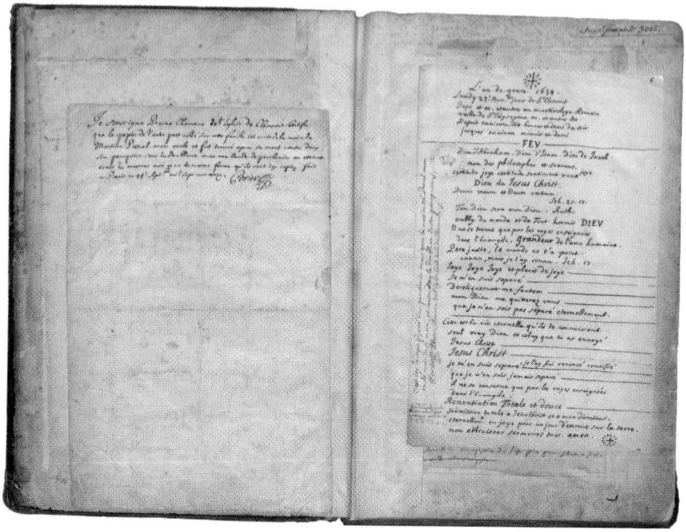

pocket of inner pleat, finding and then extracting a folded parchment wrapper containing the paper on which Pascal had scribbled his words (illus. 2). He noted that on the inside of the parchment wrapper there was copied, in elegant calligraphy, a second and neater version of the original text. Pascal's nephew, Louis Périer, then the canon of the church of Clermont, reported having produced in 1692 a 'figured copy' of the text written on the parchment (illus. 3). In 1711 Périer glued the paper and parchment onto backing board, and in 1732 Guerrier, who had never seen the parchment itself, brought the discovery to light. Legend has it that at the time of the demolition of Port-Royal the parchment copy might have been cut into pieces and distributed to the faithful.[3] Whatever its fate, it can be imagined that over the last eight years of his life Pascal (who died on 19 August 1662) kept close to his beating heart, in secret, his autograph of the illumination.

3 Pascal's Memorial, 'figured copy'.

Although the fortunes of the paper and its copied wrapper have the trappings of a mystical fable, the aspect of the document calls attention to what makes Pascal's writings at once illuminating and mysterious. In the paper version a modest and seemingly hastily drawn cross (the vertical stroke bloated by a drop of ink at the juncture with the horizontal) stands over the first line that dates the event, 'L'an de grâce 1654', and four lines below, isolated, 'feu' (fire) is penned in minuscule. The 'figured copy' is coiffed by a carefully traced cross from which radiate eleven dashes. Below, in cursive, five lines of text lead to an epigraphic 'FEV', in majuscule, which two lines on either side of the word emphasize by extending to the margins. The reader's eye catches salient words below ('Dieu', 'grandeur', 'Jesus Christ') before the text ends on *amen* at the bottom of the page, adjacent to another cross, also with eleven radii. Readers have noted that text itself is a list of quotations, a collection of biblical references, but also, as if in anticipation of Rimbaud's famous aphorism, 'Je est un autre' (I is an other), a testament to something or someone else who is the author of the illumination. Be they schoolchildren of days past, carrying in their backpacks musty copies of the *Pensées* in timeless 'Classiques Larousse', or starry-eyed readers of the Pléiade or other critical editions, readers cannot forget the unsettling sight of the Memorial.

It may be that between the original and the copy the two instances of the sign of the cross and the word 'feu' are emblems of the *Pensées* as a whole. In her text Caws suggests why. In her discussion of the 'two infinites' (a capital moment in Pascal's notes for what would have been an dialogic apology for the Christian religion), if only to draw attention to the mix of gravity and searing wit in Pascal's French, she writes that everywhere translation of the *Pensées* remains a thorny enterprise. Rehearsing the staging of the wager where the believer seeks to convince the agnostic or libertine of the existence of God, she notes how for Pascal reason 'cannot determine the question of God, but it is rational to make some attempt at it'. The person whom Pascal begs to flip a coin to ascertain yes or no indeed faces a win–win outcome: if the coin drops heads, yes, God exists. If tails, the very doubt that comes with the fall of the coin can be taken as a symptom or sign of his presence. Further, because the bet is – and must be – undertaken, the wagerer is embarked, off to sea as it were, floating rudderless on an infinite ocean. Reflecting on this text, Caws notes that the word Pascal uses for 'heads', *croix*, 'which is translated as "face", refers to the heads portion of a coin, but of course *une croix* is also a cross, so that the symbol enters into the wager. Now, of course, here as elsewhere, translation matters enormously.'

And so it does, and even more if the symbol or 'figure' that marked the Memorial can be taken literally, at face value, in the drift of the spoken register of the words. The silly sign at the head of the Memorial (what in the copy also appears at the bottom) could be an imperative to believe: when uttered, *croix* (both heads and cross) carries its homonym, *crois* (I/you believe), an imperative to avow or to succumb to faith. Like a double entendre in an advertiser's carefully crafted slogan, the amphibology becomes an expression or prompt of doubt, in other words, a sign of a secret or, in the visual regime of the Memorial, hidden or concealed evidence of God. Caws brackets the wit – the disarming simplicity, even the riotous stupidity

– of Pascal's manner when she is compelled to translate a sentence from the passage on the wager. 'Let us weigh the gain and the loss in calling heads that God exists (*croix que Dieu est* [heads/believe that God is]). Let us assess the two cases. If you win, you win everything; if you lose, you lose nothing. Wager then, without hesitation that he exists.' 'Croix que Dieu est' enters the reading in the mode of a voice-off in a movie, a statement stated out of frame, perhaps also a fragment of indirect discourse, or even in what Roland Barthes noted when observing that comments inserted between parentheses in a discourse generally belong to a different speaking, writing or even whispering subject (to an other). Already the utterance of 'Dieu est', the fragment by which we are told to believe, almost invites the happy misprision where, heard in his name, *tu* shares the place of *Dieu*. 'Dieu est . . . tu es' (God is, you are), and by extension, 'il est' (he is): thus the speaker conjugates and in conjugating is conjugated with God.

It is here where Caws gives us reason and rationale to consider at greater length the stakes of a 'manner of writing' (*manière d'écrire*), laced and riddled with secrets, where dazzling twists and turns are taken taken between what is seen and what is said. The style is the subject of an essay by Michel de Certeau on Pascal's mystical manner. In a close and protracted reading of the Fourth Letter Pascal addresses to Charlotte de Roannez (then aged 23, following her conversion in the church of Port-Royal on 4 August 1656 and before she entered the abbey to be received as a novice), Certeau finds that as a whole the 'secret and the visibility of these letters grow together', and that the passing of time has turned the personal correspondence into 'pensées', 'stones whence the singularity of the events are effaced', causing the text to become visible exactly where its author disappears.[4] The events or 'experiences' gained from the reading bring forward 'what Pascal calls a "manner of writing" and the ways of "turning things" [*tourner les choses*]. "The ways of turning a single thing are infinite."'[5] Citing the *Pensées* where Pascal notes 'how statements can be varied' and how 'propositions can be shifted in every direction', Certeau argues for what he calls transformational practices. 'Manners [or styles] of speaking and manners [or styles] of doing produce a new discourse even if the matter is ancient [*les matières sont anciennes*] . . . "Turns" multiply possibilities, they set things in movement, they cause them to turn and come back to the instability of their equilibriums.'[6] He writes that these turns 'have the mobile rigour of a style'.[7] At once obvious and concealed in the Fourth Letter, they mark what Pascal calls this

> strange secret of God's withdrawal from and concealment in the sentient world. Impenetrable to the sight of humans, the secret has remained hidden under the veil of nature . . . It [or he] had been far more recognizable when invisible, and not when made visible . . . All things [*Toutes choses*] are veils that cover God . . . May we render to him infinite graces of what, being veiled in all things for others, is discovered in all things and for us in so many manners [*il s'est découvert en toutes choses et tant de manières pour nous*].[8]

Following close analysis of the letter's speech-acts, Certeau shows how the words are tactically drawn, how any sense of certitude is best stimulated when simulated, and above all how something 'goes by' [*passe*] when its 'incisive rapidity and penetration' change the position of the reader before he or she realizes that an alteration has taken place. 'Through required passage, via broken connections, logical jumps or accelerated ellipses that spur movement, Pascal's way of doing things [*la manière de faire pascalienne*] leaves readers with neither time to react nor to remain in place.'[9] His method aims at movement, at a discourse set in motion, becoming the product of the 'very mobility of its process', which never settles upon a given place. As he puts it in a terse formula having the ring of Montaigne's *Essais*, to 'think the secret is to learn how to read it' (*penser le secret, c'est apprendre à le lire*).[10] To think is to see, but to see only in motion because, in the identity of differences of positions sensed though mobility, things in places that are at once visible and legible conceal and reveal something else, something ineffable, that in passing – and only in passing – is apprehended and no sooner lost. 'The Pascalian gesture is written in the contents of the text. A religious field is crossed by this manner of thinking that produces an effect of a "lost God" and of a truth that 'wanders unknown among humans.'[11] As if turning and inverting the Cartesian idea that to think is to weigh, where *penser* is the doublet of *peser* Pascal suggests that to think is to pass by, *que penser, c'est passer.* He notes further that where the Old Testament is cited in the text of the Fourth Letter to Charlotte the quotation marks surrounding 'Isaiah' and others of biblical gist are virtual halos, as it were, signalling the 'distance of an echo' of an age-old truth passing through the historical landscape of the semantic field. As a result 'vision is no longer at stake but, rather, writing. Visibility, mediated by an act of believing [*un croire*], "I believe" [*je crois*], is merely the address of the other [*le lieu du dire de l'autre*], a statement addressed to the veritable Other [*dire adressé à l'Autre véritable*].'[12]

In the paragraph above, the digression by way of Certeau's reading of the 'strange secret' of the hidden god is taken to suggest that Caws's lifelong attraction to Pascal follows the visual traces of a restless and ever-mobile writing bearing on matters at once casual – trivial even – and on the surface of things profoundly unsettling. Throughout the monograph she remarks how much Montaigne gets under Pascal's skin. Referring to Pascal's admiration for the author who was 'the imcomparable author of the art of conversing [*l'incomparable auteur de l'art de conférer*]', she implies that, as it was for Certeau, the force of the *Pensées* and other writings owes to a craft, like that of great painters, of engendering an effect of movement in the difference of words and things, both visible and legible, that 'translate' into and away from each other. In 'De l'art de conférer' Montaigne invokes painting at great length, and so also the tact and 'manner' of handling the art of speech. As if furnishing a prompt for Pascal, he writes: 'because we're dealing with the manner and not the matter of speech' [*car nous sommes sur la manière et non la matière du dire*]. In passing, the formula suggests that the liminal difference between *manière* and *matière* is decisive where speech and dialogue are at issue. Pascal found the 'manner' of the *Essais* especially distasteful where, in vanity and self-conceit, his words depicted or even portrayed the writer. Paradoxically, what he called 'le sot projet qu'il a eu de se peindre' (the foolish project he had to depict himself) becomes for Pascal a negative pole of inspiration, and not in the least because where matters of knowledge and being are at stake the concealed relation that visibility holds with legibility is shaped by the poet's 'manner' of writing and thinking.

As an invitation to enjoy *Blaise Pascal* from this angle it might be worth considering one of Pascal's most celebrated *Pensées*. The *roseau pensant* or 'thinking reed', the turn of speech he coins to typify the sorry but nonetheless heroic condition of humankind, remains an arresting *figure* which, like Pascal's Memorial, everyone of French extraction carries close to their heart.[13]

'L'homme n'est qu'un roseau, mais c'est un roseau pensant' (Man is merely a reed, but it is a thinking reed). And, like the many clichés peppering the Memorial and inspiring the styles of the *Pensées*, the striking visibility of the verbal icon – a disbeliever would ask what in hell or in god's name is a *roseau pensant* – is found in the manner of its writing. It has been shown that in crafting the aphorism Pascal might have found grist for his mill in Montaigne's 'Apologie de Raimond Sebond', the epic and grimly comic demolition of human presumption and, at the same time, a searing interrogation of the existence of God, the presence of whom is discernible, if discernible at all, through a leap of faith.[14] Raging and railing against the vanity that puts humans on an upper rung of the ladder of being, Montaigne turns the world order topsy-turvy, dropping 'man' onto the dung heap of the cosmos:

> *La plus calamiteuse & ~~foible~~ fraile de toutes les creatures c'est l'homme, & quant & quant, ~~dict Pline,~~ la plus orgueilleuse. Elle se sent et se void logée icy, parmy la bourbe et le fient du monde, attachée*

> *et clouée à la pire, plus morte et croupie partie de l'univers, au dernier estage du logis et le plus esloigné de la voute celeste, avec les animaux de la pire condition des trois; et se va plantant par imagination au dessus du cercle de la Lune et ramenant le ciel soubs ses pieds.*[15]

> The most calamitous and ~~feeble~~ frail of all creatures is man, and by and by, ~~says Pliny,~~ the most boastful. It senses and sees itself lodged here, all about the mire and shit of the world, attached and nailed to the worst, the deadest and most stagnant part of the universe, at the lowest level of the dwelling and at the farthest remove from the celestial vault, with animals of the worst condition of the three; and in its imagination it goes planting itself above the orbit of the moon and bringing the sky beneath its feet.

Montaigne turns man (in the masculine) into a sad creature (in the feminine, or, in recall of Latin, in an indifferently neuter state of 'it'). Why is it calamitous? Every Latinist knows – and, like Pascal, Montaigne was an impeccable Latinist – that 'calamity' derives from *calamus*, a reed or stalk, much like what is found in a marsh or even a field of wheat. A calamity would result from a harvest of reeds whose seeds at their crowns have been blown away or subjected to rust and blight. Yet *calamus* is also related to a writing instrument. In his epochal essay on the *roseau pensant* Alfred Glauser noted that Pascal 'translates' Montaigne's description of the 'calamitous' condition of man into the figure of the reed. The human creature's redemption owes to the fact that it thinks, in other

words, *qu'il pense*. In its passing and passage, in the cinematic flicker between Latin and French, *calamus* cuts through the *roseau* that becomes an icon, at once visible and legible, of the apparatus of writing. To think is to write, and to write is to believe. In view of the 'strange secrets' of Pascal's manner and style, the reed is something of a moving image, a figure that stands high when calm, then bends back and forward under the effect of the breeze or wind, the spirit, the breath, or *spiritus* that animates all things.

If distortion and verbal anamorphosis are in play, the *roseau pensant* could also be, in passing, a *roseau penchant*, an ever-inclining or ever-bending reed that wafts in wind emitted from an unknown origin. Thus, if to think is to write, and if to write is to believe, to believe is *to be inclined*. Whenever someone avows having a penchant or is 'inclined to believe', the person would be stating a happy tautology simply because

4 Jacques Callot, *The Reeds and the Wind*, c. 1628, etching.

belief comes when thinking bends according to the ineffable measure of creation. Because the brilliance of Pascal's manner owes in part to how he bends clichés, it can be wondered if the thinking or 'penchant' reed, in addition to Montaigne's 'Apologie de Raimond Sebond', might also have an origin in a casual or everyday image, such as Jacques Callot's etching of reeds in the wind (illus. 4).

In Callot's image the hidden god would be a wind head, borrowed from the boreal region of a map, possibly a cherub or a putto, or merely an illusion of a face glimpsed in a cloud. The reflection of the arched reeds on the rippled surface of the water depicted by tight parallel hatching giving way to outlines of cloudy bubbles and suave squiggles underscores how much drawn lines are part of the process of the etching. They are its cause and effect – like Pascal's *roseau*, they are virtual writing instruments bending in the wind. They affirm that the *roseau pensant* could well be the same *roseau penchant*, the leaning reed with a penchant to incline or, implicitly, to 'be inclined' to acknowledge and thus, in pathetic fallacy, to 'believe' in the wind. For Pascal, for whom all things are living testaments to Creation, the reed's inclination or tendency to respond to the call of the wind would be evidence of its will to believe. For Callot, whom the art historian Henri Focillon long ago called one of the great French *writers*, the reed can be both pen and burin.[16] And like Pascal, in its drawing or in its visibility, its will or inclination 'to think' makes manifest a 'strange secret' of Creation.

The reader of *Blaise Pascal* is led to sense why Caws, in whose genial writings on poetry and painting from Char to Motherwell and from James to Jabès the visible and invisible

5 Georges de La Tour, *St Sebastian Attended by Ste Irene*, 1649, oil on canvas.

are of the same order, might also believe that to write is to think, and to see is to read. Hence a life lived with Pascal. Finding drive and pulsation in his words, she lives with the *Pensées*. By way of conclusion, it may have been a visible scrap, a reminder or remainder of the *Pensées* that drew her to René Char, whose *parole de fragment* (speech of fragment) inspired a common admiration of Georges de La Tour, whose great painting *St Sebastian Attended by Ste Irene* (illus. 5) ought to be read in candlelight. With her left hand Ste Irene takes the pulse of the cadaver while, with her right, she holds a cylinder at whose top a flame wafts and throbs and where, too, are concealed, yet plainly *seen* and *written*, the numbers of the Vulgate – 392 – referring to the words that describe the scene itself. The blood of the martyred saint, having succumbed to the shaft of an arrow lodged in his sculpted belly, as if passing through the living saint's hand holding the cadaver's wrist, comes to life when evanescing in the writing of the burning flame. Such, too, is the legacy of the pulse of a visibly thinking mode and manner of writing, here and elsewhere, that belongs to Caws and Caws alone.

ONE

Early Upbringing and Arrival in Paris, 1623–41

THE MEMBERS OF the Pascal family were, depending on your point of view, petits bourgeois or aristocrats of the *noblesse de robe*, from the humble provincial elite of Auvergne. The father, Étienne Pascal, was gifted in mathematics, science and ancient languages – only the last of which were really current interests of that class and that epoch. In 1608 Étienne went to Paris to study, after which he returned to Clermont-en-Auvergne and was able to buy the post of *conseiller*, a kind of magistrate in fiscal matters. At 28, in 1616, he married Antoinette Begon and they lived together in Clermont. After losing a daughter, Antonia, the couple gave birth to Gilberte, in 1620, and Blaise on 19 June 1623. When Étienne was able to purchase a more important post, the family moved to a home near the cathedral, where Jacqueline was born, on 5 October 1625.

Two wonderful and peculiar characteristics marked Blaise Pascal's first year: he could not stand to see running water, nor to see his mother and father together (especially if they were kissing). He would enter then a kind of trance and be subject to violent convulsions; he would go into this state throughout his life, at times of exceptional stress. Ah, says Jacques Attali, see how in this family everything to do with sexuality is 'hidden, masked, censored'.[1] This is worth noting from the beginning, as it explains many of the ongoing complications, particularly in relation to the brother–sister imbroglio to be discussed later, as Jacqueline and Pascal forged a bond seemingly more interwoven than that of a marriage. Perhaps there is something to Pascal's early, and quite physical, abhorrence for displays of affection. He may have already intuited what he came to believe later, after becoming a follower of Jansenius and adhering to the *Augustinus*, his treatise on St Augustine's theology: that we are born into original sin despite the chasms of time and mythology separating us from the Garden of Eden.[2] The early Church Fathers were not averse to making claims about sexuality and its purpose for procreation, not pleasure. St Justin Martyr, a second-century Christian apologist, claimed that 'we Christians either marry only to produce children, or, if we refuse to marry, are completely continent.'[3] Other second-century Christian moralists, such as Clement of Alexandria, declared: 'husbands [are] to use their wives moderately and only for the raising up of children', pronouncing that 'to have coition other than to procreate children is to do injury to nature.'[4] Pope Damasus I, an early expounder of the Christian religion, whose acolyte, St Jerome, was the translator of the Vulgate Bible, postulated and preached that 'Adam and Eve were created sexless; their sin in Eden led to the horrors of sexual reproduction. If only our earliest progenitors had obeyed God, we would be procreating less sinfully.'[5] In the *Pensées*, when Pascal addresses 'les divertissements', those distractions that pull us further from God and deeper into sin, sexual promiscuity is lumped in with other diversions, such as gaming, drunkenness and prostitution. There is little

6 François II Quesnel, *Blaise Pascal*, *c.* 1690, oil on canvas.

7 *Gilberte Périer*, 17th century.

question that Pascal, even from an early age, understood what it was to consecrate the body to God, if only instinctively. His maladies and hysterics, therefore, may have been the indication of something greater stirring within.

Étienne Pascal may have in fact believed that his son's ailments arose from a sorceress having cursed him in his crib. Such a tale can be pieced together from Blaise's sister Gilberte's *Vie de Pascal* (illus. 7). About this incident we do not have many specifics, though one interpretation goes that Étienne, a man of science, was so frustrated with the medical professionals who could not help his son that he abandoned reason and sought a cure elsewhere. He allowed himself to believe that if a sorceress had done this to his child, she could undo it just the same – an indication, perhaps, of how much the father adored his son.[6] In the seventeenth century, despite its scientific and religious authorities, there was still very much a belief in magic and the casting of spells. An old woman was accused of cursing the infant Blaise because she uttered a spell at Pascal's mother, Antoinette, who refused to give her alms; the woman's witchery is questionable, however, since she was threatened with being burned at the stake if she refused to confess her sorcery. So, under her guidance, Antoinette and Étienne transferred his fate to a black cat. The transferral did not work: the cat died and Blaise got no better. The sorceress then made a poultice of nine leaves from three different plants, and had a seven-year-old child cook it beneath a full moon.[7] Blaise went into a coma. His father hit the old woman, accusing her of killing his son, and the family went into fits of mourning. But then the child woke up and smiled, and was cured – the proof of which was, incontrovertibly, that he no longer reacted either to running water or to the sight of his parents embracing. Cured under suspicious circumstances, Pascal would later write in his *Pensées* about the relinquishing of reason, maybe with his father's superstitious leap in mind, claiming that 'reason's final step is to recognize that there is an infinity of things beyond it. It is merely feeble if it does not go as far as realizing this.'[8]

Pascal would be subject to convulsions all his life, and the epoch seems to have been given to superstitious behaviour and belief. When his mother died in 1626, the reason popularly given was that she had eaten too many cherries.[9] Assuming this lay diagnosis is based on some sort of seventeenth-century mystery about the effects of overeating fruit or the properties of cherries, the fruit is symbolic nevertheless. In many paintings from the century, the cherry was a mark of fertility and youth, used to show the playfulness and abundance of the maternal figure in family portraits. But we also see depictions of the infant Christ holding a pomegranate or cherries to emphasize both his sweetness and his forthcoming blood sacrifice.[10] Antoinette Pascal's death seems to have left its mark on the young family, since the widower Étienne, only 38 years old at the time, did not remarry but hired a governess, Louise Delfaut, to act as surrogate mother to his three young children, raising them and staying with the family until her death in 1658. In any case, it was Étienne who brought up and educated the children, refusing to send them to school after experiencing the insufficiencies in his own education at good regional schools and the Paris university. He wanted his children to experience higher learning from an advanced pedagogue who could teach them using curated methods extracted from the works of Rabelais and Montaigne. Jacqueline and Blaise learned French with the use of cardboard letters, and acquired knowledge of grammar through conversation rather than by memorizing arbitrary rules. Blaise even began to learn Latin at the age of about seven – before completely learning French – since he showed an aptitude for study in general, early on.

Étienne's method was to teach his children without forcing them in any way, and to explain the purpose of each subject before introducing it; he also talked to them often about nature. Gilberte described Blaise as always wanting to know the reason for every single thing, appearing more curious than Jacqueline and far more than Gilberte herself, whom she presented as the least brilliant of the three children. Perhaps she was correct, and not simply modest, but we are fortunate to have her written memoirs, which relate their lives so devotedly. As for the children's faith, instilled in them from a young age, that too may be attributed to Étienne, who, as Attali tells us, made certain to keep their education separate from religion, since faith was not a science to be taught to the mind but is a deeper knowledge already found in the heart.

PARIS AND SCIENCE

In 1631 the family moved to Paris, so Étienne could take part in scientific circles. Two years earlier, Cardinal Richelieu had been named the principal minister of the state, which meant this was not only the epoch of the academies concerned with criticism and analysis of cultural production, but also of a great disquiet. Aged eight when the family settled in the quartier du Temple, near the Marais district, Blaise was already a budding autodidact, eager to be the first to discover things, and shouting, it was said, like Christopher Columbus discovering America. As his education evolved, his father continued to instruct him, but their relationship was not particularly close, no tenderness being in evidence. But Étienne did his best to cultivate his child, conversing with him in Latin every Monday, Tuesday, Thursday and Friday. He would emphasize the importance of listening to conversations over reading books, so that Blaise could learn to think and reason for himself, and it was not long before Blaise acquired Greek, and he and his father discussed logic, physics, philosophy and astronomy. But Étienne struggled with how to introduce Pascal to mathematics without its difficult concepts overwhelming him, and so kept putting it off. Attali, however, claims that in secret, hiding it from his father, Pascal discovered at an early age the discipline that was to bring out his genius; 'c'est un coup de foudre' (it was love at first sight).[11]

In 1634, when the family had moved to the rue du Cloître-Saint-Merri, the Duc de Roannez (Artus Gouffier), living across from them, became Étienne's friend; he introduced him to a group which gathered around Marin Mersenne, an ordained priest who was connected to the scientific community and noted for his contributions in mathematics (the Mersenne prime); he has been called 'the center of the world of science and mathematics during the first half of the 1600s'.[12] Mersenne was already by then dressed in the black habit of the Minim order, and was corresponding with René Descartes and Pierre de Fermat after founding the Academia Parisiensis, a meeting of great minds who would share and discuss scientific and mathematical discoveries openly. Around this time, the eleven-year-old Pascal was experimenting on his own, showing his early curiosity and inventiveness. He clanged a knife against the glasses on the table, trying to understand about the sound, and wrote – according to the family legend – his first text, a *Treatise on Sounds*, though no trace survives. He also demonstrated the propositions of Euclid's *Elements*, using the mathematician's terms 'axioms' and 'definitions', which he may have got from a Latin text of the *Elements* he found in his father's library, tearing out the only page he could before being discovered.[13] Once Étienne had uncovered Pascal's secret and realized his son's intuitive genius for Euclidean geometry, he asked Mersenne's permission to bring him to their meetings, and the savant welcomed the boy with open arms, inducting him into a world of erudition that fostered the child's genius. So persuasive was his prodigious understanding of geometry that, at twelve and a half, Blaise was now meeting with various mathematicians at Mersenne's gatherings, discussing Epicurus, the Greek philosopher who believed both the body and soul die at the end of one's life and therefore death is nothing to fear, with Pierre Gassendi, a French philosopher, observational scientist, astronomer and mathematician whose philosophy found a balance between scepticism and dogmatism. He was also privy to conversations between mathematical greats such as Descartes and Fermat, who argued about analytic geometry in the spirit of Mersenne's academy. In the colourful description by Dominique Descotes in *Blaise Pascal: littérature et géométrie*, Père Mersenne, wanting to persuade people of the usefulness of science, invoked what he called a 'joyous arithmetic', playing on the squaring of the circle, as well as the application of geometric proportion to moral actions, on the way the infinite numbers would lead to such strange and irrational propositions, and on tricks of increasing strength.[14] From this, Pascal's perspicacity permitted him to figure out the cycloid, 'the simple roulette, and the roulettes both elongated and shortened, varying according to the relationship of the wheel and the generating circle [*le cercle générateur*]'.[15] In his *Pensées*, Pascal, perhaps reminiscing about the conversations that inspired him in his pursuit of mathematics, claims that 'there are some who speak well but do not write well . . . because the place and audience warm them up and elicit from their minds more than they can manage without that warmth.'[16] So it was always human inventiveness (*ingenium*) that instituted progress.

AUVERGNE AND ÉTIENNE'S PARDON

In 1638 the French government, financially stricken by the effects of the Thirty Years War – a succession of wars that started in 1618 as religious disputes between Catholics and Protestants and evolved into wars of rivalry for European political power – defaulted on its bonds and was unable to continue paying out annuities to its investors. This included Étienne Pascal, who had put most of his family's inheritance into bonds. The timing of the financial debacle could not have been worse, as it was during the Fronde (1648–53), a time when the French nobility were rebelling against the monarchy, essentially challenging Louis XIV's authority but ultimately strengthening it. Since Étienne was involved in the protests about finances at such a volatile time, his name was recorded and he was identified as one of the leaders of the demonstrations. For fear of imprisonment, he fled to Auvergne and began life anew in the countryside. His governmental rebellion did not haunt him for long, however, thanks to the talent of Jacqueline, then aged thirteen, who was a poet and also an actress and who, at this young age, would arrange – through the encouragement of Cardinal Richelieu, who greatly respected her artistry – to obtain her father's amnesty (illus. 8). It was in fact her role as Cassandre in Georges de Scudéry's *L'Amour tyrannique*, played in the presence of Richelieu and his niece the Duchesse d'Aiguillon, that managed to secure the pardon for her father, who was then accorded the post of deputy commissioner for taxes in Rouen. Jacqueline remained a respected poet until her entry in 1652 into Port-Royal L'abbaye in Paris (not the associated Port-Royal des Champs in the country), when she was told to give up the creative life, which she did.

In 1639, at the age of sixteen, Pascal presented his 'Essay on Conic Sections' (which contained a theorem, now called 'Pascal's theorem' or 'Hexagrammum mysticum theorem') at a session chaired by Mersenne and attended by many well-known mathematicians, such as Girard Desargues, the French mathematician and engineer who formulated a theory of involution, of triangular perspective and – most important for Pascal – a unified theory of conic sections, inspiring Pascal's work on the same. No one could perceive how crucially important Pascal's foundational work on conics would be to later studies in projective geometry, especially those taken up in the nineteenth and twentieth centuries – modern architecture and industrial design included.[17] Projective geometry differs from Euclidean geometry in that it does not only reveal shapes as they are but describes objects as they appear.

The difference is that, since the length and angles of objects may be distorted when we look at them, projective geometry offers a way of seeing the object as it appears on a three-dimensional plane. Writing about 'Pascal's Physics', Daniel C. Fouke describes this notion, which enables our point of

8 Philippe de Champaigne, *Cardinal de Richelieu*, 1633–40, oil on canvas.

view or perspective to discover unities that do not appear upon first view:

> Imagining an eye perched at the summit of a cone discloses that points, a straight line, a circle, an ellipse, a parabola . . . are images of one another. Taking a circle as the image that is projected, circumscribing figures around or inscribing figures within the circle and projecting the points from those figures onto the other sections of the cone reveals their common properties.

9 Portrait sketch of Pascal as an adolescent by his friend Jean Domat, *c.* 1649.

> The simplicity and symmetry of the circle makes it easy to grasp its properties, which when projected on to the other conic sections, enable one to unlock the properties of the cone.[18]

Pascal's theorem in particular posits that if six arbitrary points of a conic section are joined by line segments to form a hexagon, the three pairs of opposite sides of the hexagon will meet in three points lying on a straight line (the 'Pascal line'). His theorem is recognized as a generalization of the hexagon theorem of Pappus of Alexandria, a fourth-century Greek mathematician. Though Pascal did not include a proof with his theorem, it was quickly taken up by others and, as is always the case with Pascal's fevered erudition, proven and expanded beyond the foundation of its genius. We might recall Albert Béguin here: '[Pascal was] distracted from a work as soon as he had understood its first principles, as if he wanted to leave to others, who had more leisure, all the concern of inquiries and verifications.'[19]

When Pascal demonstrated his findings, he made a big impression, leading one of the auditors to say he produced more corollaries 'than does a plum tree plums. You'd just shake him a little and they pour down around him.'[20] Also impressed with the young mathematician, Mersenne claimed that Pascal had 'crushed everyone who had ever treated the subject',[21] and Desargues termed the property of the hexagon 'la Pascale', in a professional admiration that was reciprocated by the budding genius. But Descartes' jealousy of Pascal showed itself at this point and he accused the sixteen-year-old of plagiarizing Desargues, an accusation that was always to undermine the relationship between Descartes and Pascal, though Pascal's regard for his geometry mentor would not dampen, as he paid homage in his *Discours sur la religion* to 'Desargues, Lyonnais, one of the great minds of this time and one of the most learned in mathematics, and in cones'.[22]

PERPETUAL MALADIES AND FAMILY ADOPTIONS

Pascal's creative fury and mathematical inventiveness could not assuage his maladies, however, and he remained exceedingly nervous, suffering from migraines and even paralysis. He was nevertheless supported by the small group of admirers around him, with a loving friendly relationship with Artus Gouffier, Duc de Roannez, and his sister Charlotte, along with his own brilliant sister Jacqueline, who wrote poetry and learned entire plays by heart. She herself was stricken by smallpox and retained a few marks on her cheeks ever after, perhaps contributing to her vulnerability and unwavering compassion for her brother's fragility, and maybe even her desire for refuge in the convent at Port-Royal.

In 1641 Florin Périer, Étienne's young cousin who was living with the family in Rouen, married Gilberte, and the couple remained in the family home. Blaise, suffering excruciating headaches and stomachaches, was paralysed at this time from the waist down, and used crutches from then on, whenever this paralysis recurred – which seems to have been frequently. Generally, he nourished himself only on liquids, solids being problematic for his stomach. Gilberte reported later that from the age of eighteen, Blaise never spent a day without pain. Of the doctors who tried to cure him, he declared that they had only 'imaginary science' and that their medical costumes permitted them to dupe the world – a story reminiscent of the plot of Molière's *Le Malade imaginaire* (The Imaginary Invalid), with its hypochondriac, Argan, battling tête-à-tête with his physician, M Diafoirus. Jacques Attali, the most dramatic of Pascal's biographers, compares this excess malady to his similar behaviour at the age of one, already described, when his parents embraced each other: as soon as a family member was occupied with someone other than him, Pascal's body would react violently.[23] But as we will see in the following chapter, Rouen was still a place of creative genius, where he came up with his 'arithmetic machine' to help his father, as well as a scene of spiritual awakening: it was where he experienced his first conversion, one that set him on a course from which he would never waver.

TWO

Rouen and the Pascaline; Science Trials and Conceits; the 'First Conversion'

BY THE END OF 1642 Cardinal Richelieu had died and the precarious situation of the French government, as well as the Pascals' relation to it, set the family on shaky ground. As their alliance with Richelieu came to its end, they were uncertain how to garner the same favour with his acolyte and successor, Cardinal Mazarin, an Italian diplomat and politician. Mazarin was appointed Chief Minister of France upon Richelieu's death, and the Queen Regent, Anne of Austria, gave him even more power in 1643 when Louis XIII died, leaving behind a five-year-old heir too young to rule. With France still at war with Spain, the political mood was more than uneasy. Florin and Gilberte Périer had their first son around this time, and the new family addition seemed to affect Blaise very little, his mysterious ailments holding steady. When his brother-in-law left his post as assistant to Étienne Pascal in his position as a tax collector in Rouen, one of Richelieu's most gainful states, he and his wife moved back to Auvergne, where he returned to a position in Clermont's parliament. For a short time, they left their young son, Étienne Périer, in the care of his grandfather, hoping for him to benefit from the same curated education the three siblings had received. Blaise was distracted by the child, and drew closer to Jacqueline, finding solace in his younger sister's company – circumstances that may have set the stage for the bond they would forge later on. With Florin gone, Étienne Pascal was left to do a mountain of work on his own; this was perhaps an inspiration for his son, who developed his 'machine à penser' (thinking machine) around this time, in the hope of automating tax calculation.

THE ARITHMETIC MACHINE

Pascal was only nineteen when he constructed his arithmetic machine to do addition and subtraction, as well as multiplication and division, based on a system of clocks. His invention, designed to assist his father with his accounting, resembled the device used in offices from the 1850s onward. The machine, eventually called a Pascaline, was another example of his inventiveness in solving a problem by relying on the foundation of his education: to understand how things worked. Not only was he trying to devise a machine to help his father count and collect taxes, but he may have been improving on the device of Wilhelm Schickard, a professor of Hebrew and astronomy at the University of Tübingen from 1619 until his death in 1635, who had invented a calculating clock based on Napier's bones, a manually operated counting device.[1] In Schickard's surviving notes on the device, we learn that it may have malfunctioned or been incomplete.[2] Pascal would have no such problem with his *machine à penser*, however, and he worked on it for the next two to three years, while also replacing

Florin as his father's assistant, helping him with the daily fiscal administration.[3]

When the machine was ready to be sold to the public, Pascal emphasized in his attached description that he could not explain how the complicated device functioned in writing, but that those interested might purchase one, as well as receive an explanation of its operation, from his respectable friend Gilles Roberval, professor of mathematics at the Collège Royal (illus. 10). He also assured those curious that the contraption was too complex to emulate properly and that to buy an imitation would be a grave mistake, since even the most talented and skilled artisan could not replicate the instrument

Photo Panajou.

1. Machine arithmétique (fermée).

2. Machine arithmétique (ouverte).

10 Pascal's calculating machine.

without Pascal's guidance and precise measurements. He confessed to having made more than fifty prototypes, using one of them successfully for years for his father's municipal tax collection. The significance of his contribution explains his pride in the instrument as well as his fierce defence of its originality. In a letter dedicated to the chancellor of France, Pierre Séguier, Pascal requested his 'glorious protection', since 'unknown inventions always have more censors than people who approve', seeming to underline his fear that people would assume it impossible for him to have come up with it on his own since it was too complicated for them to understand.[4] His handwritten inscription on the inside cover of the oldest of the seven machines made in France during his lifetime (the high price, as well as Jean François' negative report about the machine in his book *L'Arithmétique* in 1653, seems to have deterred sales) reads thus: *Esto probati instrumenti symbolum hoc. Blasius Pascal arvernus, inventor. 20 mai 1652* (This symbol is a stamp of approval. Blaise Pascal of Auvergne inventor. 20 May 1652).[5] And his pride in his invention continued in 1652, when he gifted one of his machines to Christina, Queen of Sweden, who was reputed for her erudition, religious fervour and admiration of scientists and mathematicians. His letter invokes her estimable union of sovereign authority and respect for science, shown in both the way she ruled and the admiration she held for the scientific community. Pascal's passion for his invention as well as his satisfaction with his contribution come across in his opening words to her, which claim that both his intellectual zeal for, and dedication to, the project could not be greater if he had planned for it to be presented to the queen herself.[6]

His self-esteem is warranted, since he was the first to manipulate figures and signs automatically, and no other machine of the same sort, except for Schickard's calculating clock, would be produced until Charles Babbage's mechanical computers, the Difference Engine of 1822 and the Analytical Engine of 1837.[7] Modern-day computers followed from this, when the decimal system passed over into the binary one. In contemporary times, we may gain insight into this kind of calculating and decoding system through the work of the brilliant mathematician and cryptographer Alan Turing. The Turing machine was the origin of our computers, and, as we have seen, Pascal's arithmetic machine, the Pascaline, was at the origin of that origin, the ancestor of it all.

Pascal's life in Rouen, a small town in Upper Normandy along the Seine, seems in fact to have inspired his swells of ingenuity, coming and going like the tides. As Pascal claimed in his *Pensées*, 'Nature acts by progress, *back and forth*. It goes and returns, then advances farther, then regresses twice as much, then moves more forward than ever, etc.'[8] And thus everyone benefits from this cyclical nature of invention and the treasures its ebbs and flows wash in.

SCIENTIFIC VIGOUR

This was also an important moment for Pascal's scientific work. It is interesting to contemplate his working methods as Albert Béguin refers to them. About his manner of working as associated with his personality, Béguin offers an interesting reflection. As opposed to those who mature progressively, Pascal, sickly from childhood and conscious of the probable brevity of his life, used the only resources he had: those of his youth. As Béguin puts it,

> But Pascal, his whole style of life and of thought show how he used his present gifts, not knowing how long he had. Thus this haste with which he threw himself into his work, this suddenness of his approach, and also this impatience, always in motion, which distracted him from a work as soon as he had understood its first principles.[9]

As mentioned in the previous chapter, the young Pascal wrote, in 1639, his 'Essay on Conic Sections', based on his study of Girard Desargues' work on synthetic projective geometry. But Pascal's experimental work was not limited to geometry and mathematics. His contribution to the study of hydrodynamics and hydrostatics, centring on the principles of hydraulic fluids, was in fact just as impressive. In 1646, Pascal proved that hydrostatic pressure depends not on the weight of the fluid but on the difference in elevation, demonstrating this principle by placing a long, thin tube (10 metres) into a barrel full of water and then also filling the tube with water. The increase in hydrostatic pressure, 'the pressure exerted by a fluid at equilibrium at a given point within the fluid, due to the force of gravity',[10] caused the barrel to burst, in what became known as the 'Pascal's barrel' experiment. In another spate of scientific trials, he uncovered the principle known as Pascal's law, which proved that fluid pressure remains constant throughout a closed system. This particular finding led to his invention of a syringe used in the testing of the transmission of fluid pressure, as well as to Joseph Bramah's invention of the hydraulic press in 1795, which uses hydraulic pressure to multiply force.

By 1646, Pascal had also learned of Evangelista Torricelli, a Florentine physicist and mathematician whose famous experiment with mercury in a glass tube sitting in a basin of the same quicksilver – which essentially led to the discovery of the barometer – seems to have inspired Pascal's foray into the study of atmospheric pressure. Through his own scientific prowess and curiosity, Pascal wondered about the force keeping the mercury in the tube. As Ben Rogers claims, Torricelli 'contended that space was indeed empty, but orthodox scholastic thinkers taught, as a mainstay of scholastic science, to believe that "nature abhors a vacuum," disagreed'.[11] Pascal set out to prove Torricelli was in fact correct, delving into his own experiments, 'a series of extraordinarily elaborate and rigorous investigations stretching over four years'.[12] In 1647 Pascal produced his first tract on the problem, *Expériences nouvelles touchant le vide* (New Experiments Involving the Vacuum), whose notice to the reader emphasized his reluctance to give an entire treatise detailing the new experiments he had essayed to prove the existence of a vacuum, as well as a full report of the findings his experiments had yielded. He would rather, he writes, the reader know the abstract's core in advance so that he might see the purpose of the whole work first,[13] assuring his reader that his abstract refers to *his* experiments only, not those of another, since only his will show his genius.[14] Despite Pascal's interest in detailing only the core of his abstract, he lists the instruments he used for his experiments, such as pipes, syringes, bellows and siphons of differing lengths and

11 Pascal portrait, 1839.

shapes, along with the substances, such as quicksilver, water, wine, oil and air, claiming his tract shows how he was able to remove every material known to nature and the senses from the biggest vessel he could make, as well as how much force was needed to make it.[15] Essentially, he states that he has invalidated the ancient belief, as passed down from Aristotle, that a void cannot exist in nature since, according to Aristotle, something more dense will fill the empty spot.[16]

What seems most interesting about Pascal's scientific fervour and desire for tangible proof is the conflict it presents for his admiration of miracles. His need for evidence drove his scientific discoveries, and yet his unshakeable faith in things unseen was a marker of his religious enthusiasm and austerity. In his *Pensées*, he refers to miracles as the strongest proof for God's existence, using John 10:38, which says, 'Though you do not believe me, believe at least the miracles.'[17] As Desmond Clarke suggests, for Pascal, 'the intensity of his personal faith and his public commitment to the rigorous piety of Jansenism made it impossible for him not to reflect on the status of scientific results that were confirmed by what appeared to be incontrovertible experimental evidence.'[18] His spirit of scientific inquiry, however, never faltered and he developed his own experiment to disprove the assertion that some invisible matter was holding the mercury in the tube. In autumn of 1648, convinced he needed further proof to support his theory on the void, Pascal enlisted his brother-in-law, Florin Périer, to carry out the Puy de Dôme experiment, which his maladies were preventing him from doing himself. The vigorous climb up the mountain and the results it yielded are preserved in Florin's letter to Pascal, dated 22 September 1648.[19] In it, he describes the experiment in detail, giving specifics about his process, including measurements and distances, but he also lists the short biographies of those respectable and trustworthy people who helped him with the task. Florin is careful to mention that his assistants are both ecclesiastics and laypersons, suggesting a reliable and unbiased party in his service. The letter itself is a lovely artefact from the scientific life of Pascal, and it also reveals the reverence he elicited from his family members, especially with regard to his scientific involvements. Florin opens his letter with a sincere apology at not having got around to the experiment sooner, with weather and responsibilities getting in the way, and then expresses a hope that his full and faithful account of the event will show Pascal the precision and care with which he undertook the task.[20]

The experiment was successful, for Pascal's most astounding realization, as Attali puts it, was that the weight of air changes with altitude. In other words, the air at the top of a mountain is thinner than that on the ground, which means the measure of mercury in the barometer will differ at the mountain's peak. Considering what was known about the earth's atmosphere and the nature of our environment at the time, Pascal's hypothesis about the altitude's change in the air made formidable progress in the perception of our world.[21] Still, despite his own genius, Pascal would accredit the theories and trials of those who came before him with helping him reach his conclusion; he writes in *Discours sur la religion et sur quelques autres sujets*: 'All the men who follow each other in the course of centuries have to be considered as one man existing and continuing to learn.'[22]

DOUBTERS AND CONCEITS

Once Étienne Pascal had moved his family back to Paris in 1650, Pascal was interested in developing his *Traité du vide*, wanting to propose a theory based on Florin's Puy de Dôme trial, as well as the one he is believed to have carried out at the bell tower of the church of Saint-Jacques-de-la-Boucherie. As was often the case, he was challenged by others who refused to accept his findings. It seems Pascal was always prepared for such scepticism, as his notices and letters about his inventions attest, his parenthetical remarks often serving as pre-emptive strikes against accusations and the envy voiced by others. As we may recall, his letter to the chancellor asks for his support against those who may doubt his arithmetic machine and think it too complicated for his having constructed it. A Jesuit mathematician and professor of Descartes', Étienne Noël, in fact criticized Pascal's assertion on the vacuum after reading *Expériences nouvelles touchant le vide*, claiming that it cannot be empty space above the mercury and that it must be a substance, though of which kind he could not confirm. Noël's reasoning followed the example that despite our calling what flows through our veins blood, it is mixed with bile, phlegm and melancholy (relating to the cardinal humours, or fluids, believed at that time to be in the body), and thus so too for air, which is made up of fire, water and earth.[23] He conjectured, therefore, that the substance that enters the glass tube through its little pores to replace the mercury is purified air, separated from the rest of the air by the weight of the quicksilver descending and pulling it along.[24] He directly challenged two of Pascal's claims: that the empty space above the mercury is not filled with a pure, thinner air; and that the empty space is not filled with any material known to nature or the senses. Pascal responded with a letter of his own that not only addresses each of Noël's claims but reminds him that his theory is based on assumption and not scientific proof, since he admits to thinking about the vacuum and not actually proving that the thinner air of which he speaks exists. Pascal, the rigorous genius, is credited here with making one of the finest statements on the scientific method when he claims that for a hypothesis to be evident, not all phenomena must follow from it, but the negation of a single phenomenon will prove the falsity of the entire theory.[25]

One of the best descriptions of Pascal's experimental work is Jean Khalfa's 'Pascal's Theory of Knowledge', in which he shows how all of Pascal's scientific observations are based on 'the construction of conditions for meaningful observations, [since, as Pascal wrote,] "experiments are the true masters that we must follow in physics."'[26] Pascal's contribution to the philosophy of experience cannot be overemphasized, and as Daniel C. Fouke claims, instead of making his experiments singular and particular, he linked them as common phenomena, and though he did not discover complete new truths, he 'commandingly synthesised isolated pieces of existing knowledge . . . Pascal's experiments, whether real or imagined, build upon another, like chains of reasoning'.[27]

By this time, René Descartes had already shown his disapproval of Pascal when he accused him of plagiarizing Desargues' work on projective geometry. In a letter to Marin Mersenne dated 1 April 1640, he wrote that 'before even reading half of [*Essai pour les coniques*], [he] guessed the author had learned from Desargues, which was undeniable after his own confession to it.'[28] He goes on to demean Pascal, then only sixteen, by adding that there are many more things to say about cones which a child of sixteen would have a difficult time unravelling.[29] To understand Descartes' reaction to Pascal's discoveries with regard to the vacuum, however, we may consider how major the latter's breakthrough actually was. As Attali tells us, 'the hypothesis stating the impossibility of the vacuum is not unreasonable: the vacuum is one of the epoch's most difficult concepts to accept and the best savants, those least inclined to superstition, such as Descartes, Mersenne or Leibniz, were convinced that [the vacuum] couldn't exist.'[30] Perhaps since Aristotle, those who studied physics have tried to disprove the theory, and, as we have seen, those efforts nurtured Pascal's need to discover the evidence that negated at least one phenomenon to falsify the entire theory that nature abhors a vacuum. Thus he undertook to redo Torricelli's experiments and confirm the Florentine's findings so that he could go one step further and explain the empty space. Attali suggests that Pascal's toiling with mercury may have made him sicker than he already was, for around this time, in the spring of 1647, he suffered with greater migraines and worse stomach aches than ever before. As Gilberte put it in a letter, her brother was advised to give up his experiments for a time and find other forms of entertainment until he had rested sufficiently, which meant, for him, ordinary and worldly conversations.

And so, in the summer of 1647, Pascal was sent to Gilberte and Florin's home on rue Brisemiche in Paris to recover from his scientific activities, and it was in autumn that year that the legendary Descartes, back from his time at the court of Queen Christina of Sweden, finally met the precocious young man of whom he had heard a great deal. They actually met twice over the course of two days when Descartes visited Pascal as he lay in bed convalescing. The details of their first meeting were recorded by Jacqueline in a letter to Gilberte, dated 25 September 1647. She lets her sister know that she learned from M de Montigny, who visited their home with M Habert one evening while Pascal was at church, that Descartes expressed his eagerness to meet her brother, having always heard of the great esteem others held for Étienne Pascal and his son. Protective of her brother's fragile condition, especially in the mornings, she hesitated to agree to the meeting, but then complied, knowing the honour it would be for Pascal. The following morning Descartes arrived with a small entourage and, after a few civilities, Descartes and Pascal spoke of Pascal's arithmetic machine. The instrument was much admired when Gilles Roberval, who had also come at Pascal's request, showed it to Descartes, demonstrating it (as he was wont to do, given his task of selling the machine for Pascal). Following the demonstration, the discussion turned to Pascal's experiments on the vacuum, and Descartes, Jacqueline writes, 'with great seriousness, as we recounted an experiment and asked him what he thought had been collected in the syringe, said it was a refined material'.[31] Jacqueline does not note what Pascal answered but that he only replied as he could, ill as he was, and says that that Roberval, thinking it too difficult for Pascal to speak, chided Descartes, though civilly, who replied in return with some asperity, insisting that he would speak to Pascal as much as he liked since he was speaking reasonably. At that point, noticing it was noon, Descartes got up to leave because he was invited to dinner in Saint-Germain, as was Roberval, who shared a carriage with him. Descartes returned the following morning to finish, essentially, the visit he had begun the day before. We are not privy to their second conversation since Jacqueline was not in the room, but she does mention that Descartes stayed for several hours and that he briefly consulted with Pascal about his ailments, though 'of which he did not tell him much',[32] prescribing bed rest and plenty of bouillon until he felt better. The last detail in Jacqueline's letter to her sister which seems relevant is her mention of Pascal's letter to Mersenne, asking him to give all the reasons Descartes could bring against the column of air. Jacqueline, having read the response from Mersenne, tells Gilberte that 'it wasn't M Descartes, because, on the contrary, he strongly believes in the column of air, although his reasoning is different from my brother's, but M Roberval is the one who does not believe in it.' Mersenne also confirms Descartes' pleasure at having met Pascal and seen the arithmetic machine, but Jacqueline writes that 'we took that as simple civility.'[33]

The tension between Pascal and Descartes was anything but imagined, since letters and comments from over the years shine light on the antagonistic, patronizing and certainly competitive relationship they shared. In a letter dated two years after his visit to the ailing Pascal, Descartes claimed that he had told Pascal at their meeting to measure the height of the mercury in the tube at the top of a mountain, essentially taking credit for the Puy de Dôme experiment.[34] Since there is no record of their conversation that day, it is difficult to know the truth, but, as Attali suggests, this was most likely a lie, since Descartes' own theories on the vacuum have been shown to be too far from the truth, demonstrating the error in his thinking. He still, however, wrote two letters to Pierre de Carcavi, a fellow mathematician, in June 1649, crediting himself with the idea of the experiments Pascal eventually conducted and claiming that without his suggestion Pascal would not have thought of it, since he was of a contrary opinion on the matter.[35] But perhaps even more hurtful, if not petty and revealing of his own insecurity, Descartes wrote in a letter to Christiaan Huygens, a Dutch astronomer and physicist, shortly after his visit with Pascal, that Pascal had 'too much vacuum in his head'.

Pascal did not, it appears, appreciate the lukewarm admiration of the elder man, of whom some thought he was a disciple. Michel Le Guern suggests, however, that Pascal's acceptance of the Cartesian system was limited and exceptionable, such as when his own experience and experiments contradicted some of its parts.[36] And Pascal was not averse to sharing his opinions about Descartes openly, as can be seen, for example, by the claim that 'Monsieur Pascal called Cartesian philosophy the novel of nature, rather like the story of Don Quijote.'[37] His niece Marguerite Périer also mentions that when asked about Descartes he said: 'I can't forgive Descartes: he would have liked to eliminate God completely in his philosophy; but he couldn't help giving him a flip of a nail to set the world in motion; after that, he had nothing more to do with him.'[38]

A FIRST CONVERSION

In 1646, still in Rouen, the Pascal family was introduced to the Jansenist way of thinking. This particular event, known as the 'first conversion', is detailed in Marguerite Périer's memoir about the family. She explains how a fervent and pious priest in the nearby town of Rouville attracted parishioners from all over, two of whom, M Deslandes and M La Bouteillerie, were doctors and friends of Étienne Pascal. They decided to dedicate themselves to God; they continued to practise medicine but with a devout and charitable outlook. Pascal's father dislocated his hip when he fell on a patch of ice on his way to a duel between two men that was rumoured to be happening in a small street in Rouen. His injury was made more grave since his friends Deslandes and La Bouteillerie were out of the city at the time, and he refused the care of any other physician. When the two finally arrived to reset the leg joint, the setting proved difficult, and this called for their extended stay, since Étienne would not let anyone else tend to him. They remained with the family for three months, and Marguerite claims it was 'an opportunity birthed by God for his Providence'.[39] The two men had 'as much zeal and charity for the spiritual good of their neighbour as for the temporal, and she noticed in [her] grandfather and his entire family much spirituality'.[40] She claims that they thought it a shame that such incredible talent was spent on scientific endeavours, knowing them to be worth nothing and empty, and therefore they attached themselves to Pascal, convincing him to take up books of solid piety and give them a try. Like the other family members, Marguerite exclaims about her uncle's innate faith and his ability to see what was good and pious:

> Because he had a very solid mind and a good one, and he had never got used, even very young, to all the follies of youth, he understood from these gentlemen the good: he felt it, loved it and embraced it. And when they had won him for God, they had won the whole family.[41]

Pascal's persuasive nature and his ability to affect his entire family, including his father, is apparent in Marguerite's description of this event that changed their lives: the religious enlightenment that redirected his energy and outlook on the physical world with which he was so engaged. This was also, as we will see in the next chapter, the impetus for Jacqueline's decision to assume the veil, committing to the cloistered world of Port-Royal, sanctifying her body in both life and death, and alienating the brother to whom she had been devoted as though to a spouse.

THREE

Jacqueline Pascal, Poet and Devout of Port-Royal, 1648–52

On 7 May 1652, from behind the walls of the abbey at Port-Royal, Jacqueline Pascal wrote a long letter to her brother, wanting his approval for her betrothal to Christ and his attendance at the ceremony in which she would assume the veil and become Soeur Jacqueline de Sainte-Euphémie:

> There is no legitimate obstacle which could oppose the engagement into which I want to enter, but I still need your consent, which I am asking for with all the affection of my heart. It is not to be able to accomplish the act, because for that it is not necessary, but to accomplish it with joy, with a mind at rest, with tranquillity, because for this is is absolutely necessary and without it I will make the greatest, most glorious, and happiest action of my life with an extreme joy mingled with an extreme unhappiness, and in an agitation of mind so unworthy of this great grace that I don't believe you could be insensitive enough to cause me such a terrible sadness.[1]

Despite the picture of an unconsummated love affair between the siblings, as described by Jacques Attali,[2] we can certainly not assume anything about this being a reason for Jacqueline entering the convent. Attali writes, as always, dramatically: 'God could give her what he [Pascal] could not.'[3]

The Pascal family drama, which fills many of the biographies – Attali's account one of the more impressive of them – dwells on the intimate nature of the relationship between Pascal and his younger sister. Blaise in fact dissuaded Jacqueline from suitors by falling into trances and states of paralysis and succumbing to fainting spells whenever she seemed to be interested in someone who was requesting her hand in marriage. In 1648, when Blaise and Jacqueline were living alone with their father in the house on rue de Brisemiche, they seemed to consider themselves allied and united, as they wrote to their sister Gilberte on 1 April 1648:

> For we have to admit that it is really since this time that we should consider ourselves as really related, and that it has pleased God to join us as well in his new world by the spirit, as he had done in the earthly one by the flesh.[4]

One of the oddest elements in Pascal's life is surely his relation to his sister Jacqueline, his seemingly complete surrender to her. She was for a time in charge of their household and their life, writing and sending all their letters, even those to their sister Gilberte. On 17 October 1651, however, Blaise wrote to his older sister expressing his frustration, perhaps, at his loss of control over his own correspondence: 'I don't now know how my first letter to you ended. My sister sent it without noticing that it wasn't finished.'[5]

Pascal's loathing of sexuality, a profound disgust with the carnal, may serve as backdrop to the nature of his closeness to Jacqueline. She, having been in her own eyes the issue of the terrible sin of procreation, and he, repulsed by the very idea of anything sexual, were bound together by that kind of original shame. As Attali puts it so forthrightly:

> Blaise never got used to the idea of being born from the sexual union of his parents; he hates sexuality, even knowing that should it disappear, life itself would disappear. By his abstinence, his humility, his self-hatred, his refusal of the tenderness of others towards him, he wants to repent from being born from this monstrous action committed by his progenitor, whom he loved only because he only knew him as a widower.[6]

We know how Pascal felt about marriage from his own words on the sacrament. When a bourgeois from Auvergne proposed to Gilberte's eldest daughter, who was only fifteen at the time, Gilberte wrote to Pascal to ask his advice, claiming the marriage would be advantageous. Her brother, repulsed, it seems, by the very idea of marriage, replied: 'You cannot in any way, without hurting charity and fatally injuring your conscience and making yourself guilty of one of the gravest crimes, engage an innocent child of her age, and even of her piety, to the most perilous and base conditions of Christianity.'[7] He goes on to say that an advantageous marriage is as pleasing and desirable to the world as it is vile and detrimental in God's eyes; 'husbands, though rich and wise of the world, are in reality true pagans before God.'[8] That Blaise Pascal considered marriage 'the lowest possible condition for a Christian' certainly sums it all up.[9] His frequently expressed and absolute hatred for any human tenderness amazes and fascinates, as does this fragment, penned towards the end of his life:

> It is injust for anyone to be attached to me, even if it is done with pleasure and free will. I would surely betray anyone in whom I would incite that desire, for I am not the end of anyone and have nothing with which I could satisfy anyone.[10]

As for my personal relation to this writer and thinker, it is bound to carry in it this immanent betrayal, by the subject himself. There is something, I grant you, inherently strange in being haunted by, and particularly in writing about, someone who so adamantly rejects any such interest in himself. In this also lies the challenge of undertaking this discussion of such a self-chastising person.

JACQUELINE PASCAL, FROM POET TO SISTER

Jacqueline was a prodigy like her brother Blaise, writing poetry at eight years old and a five-act comedy at eleven. She was an acknowledged poet and, as mentioned, her acting prowess at age thirteen had persuaded Richelieu to pardon her father. She wrote poetry of political importance and it has been suggested – not in totally ironic fashion – that Pascal's own (few) poems were attempts to enter her field.

But Jacqueline gave all this up for God. Once her thoughts had turned to entering Port-Royal, she consulted its director, Mère Agnès (illus. 13), about her writing and the poetry that Richelieu had already recognized as worthwhile. She wondered if she should continue to write. Mère Agnès – who had been a votary of the original abbess Mère Angélique (illus. 12), a stalwart and strict supporter who had been

12 Philippe de Champaigne (1602–1674), *Mère Angélique Arnauld*, oil on canvas.

13 Philippe de Champaigne, *Mère Catherine-Agnès Arnauld*, 1662, oil on canvas.

named abbess at the age of eleven – consulted a priest named Antoine Singlin about Jacqueline Pascal.[11] Singlin, whom Saint-Cyran had chosen as the community's priest, 'required of its penitents absolute submission'.[12] He therefore advised Jacqueline to hide her talent, for, as Mère Agnès reports it, the female sex was supposed to indulge only in humility and silence. Though Jacqueline had waited for the death of her father before entering the convent, she never wrote another poem. Her conviction to assume the veil did not seem to waver, but her sadness at Pascal's withholding of his approval is clear in her 1652 letter to him:

> That's why I am writing to you, as the master in some sense of what must happen to me, to say to you: Don't take away from me what you aren't capable of giving me. For since God used you to procure the progress of the first movements of his grace, you know that it is from God alone that proceed all the love and joy that we have for the good and that you are able to trouble mine, but not to give me my happiness back if I come to lose it by your fault. You must know and feel somehow my tenderness by yours, and judge that if I am strong enough not to let this tenderness pass despite you, I am not strong enough perhaps to be up to the trial of pain that I will receive from such a loss.[13]

Pascal, veering back and forth in personal matters, particularly in the case of his beloved sister Jacqueline and her determination to enter the convent, gave the impression of a violently undecided brother. Hoping to dissuade her from

entering the novitiate, he had wanted to withhold her part of the inheritance from their father, hoping the convent would not accept her without a financial settlement.[14] It would take four years, Pascal said at first, but then he gave up, before changing his mind again, furious at the suggestion that it should be all to the good that it was to be decided so soon, because any delay might weaken her determination. And then

14 *Jacqueline Pascal*, c. 1600s.

once again he changed his mind. This was emotional behaviour in the extreme – just as extreme as Jacqueline's passionate engagement to Christ.

On the day of Jacqueline's formal entrance into the convent, 4 January 1652, she left the house in the early morning without warning him (illus. 14). Pascal suffered convulsions and paralysis, accompanied by extraordinary pain. Their sister Gilberte begged Jacqueline to relent, but her brother, Jacqueline said, had already stopped her leaving four times by a kind of emotional blackmail, and she was determined to succeed this time. She told Gilberte to tell Blaise that she would like to meet him in the parlour of the convent, and that he could write to her, neither of which he did. Repeatedly she wrote to him, begging him to approve her desire to enter the convent, because she did not want them to be separated emotionally.

THE EARLY PORT-ROYAL

As we know from Étienne's early conversion and that of the entire Pascal family, Jansenism was so persuasive a doctrine that it was difficult to escape it once it had entered the atmosphere. Obviously, the abbey of Port-Royal was so compelling in its firm persuasiveness that Jacqueline Pascal decided on a religious vocation to be undertaken there. (Some have suggested that this decision was a way to escape her father and brother, and as a sort of revenge on the smallpox that had disfigured her and the marriages she had not been able to make.) Ironically, the name Port-Royal originated with the idea of the swamp, or *porrois*, in which its Chevreuse building in the country was located – not the most pleasant of derivations, but it works well with the humility practised by Pascal and Jacqueline. In the ferment and intellectual movement of the seventeenth century, great writers such as Jean Racine and Blaise Pascal had close ties to Port-Royal, and the architectural simplicity of its buildings in Paris and in the country feels consistent with the liturgical bareness and the asceticism of its practice (illus. 15, 16). In similar fashion, the works of the great portraitist of the epoch, Philippe de Champaigne, are without elaboration and all the richer in their simplicity of detail, and conducive to a meditative value.[15] Pascal's relation to Port-Royal affirmed itself after Jacqueline had entered it, and he was one of the Solitaires, or bachelors, who desired to retreat to Port-Royal des Champs in the country, as he did for two weeks, a tradition begun by Antoine Le Maistre, a lawyer and nephew of Mère Angélique

15 Madeleine de Boullogne (1646–1710), *Port-Royal des Champs, Viewed from a Hill*, gouache.

who had renounced the world and retired to a small cottage beside the country abbey in 1637 where he found the 'pleasure of solitude'.[16]

But Port-Royal had been a Jansenist institution for only fifteen years before it became part of Jacqueline and Pascal's lives. Its history, in fact, is rather interesting for its quick rise as an influential haven and persuasive seat of Jansenism. In 1633, under the tutelage of Saint-Cyran, a long-time friend and follower of Cornelius Jansen, the nuns learned of Jansen's religious teachings and that one could be saved only by the touch of grace and thus be predestined for salvation. Saint-Cyran, who delivered his sermons on Sundays, took over the direction of the abbey in 1634 and instituted periods of

16 Cloisters of Port-Royal des Champs.

'renovations' – that is, separation from the world and even from the sacraments, in order to break the automatic and routine habits that stifle the imagination (illus. 18).[17] But the Jesuits, a rival group of Christians more closely tied to the Pope and staunch proponents of free will, accused him of entering the dangerous path of illuminism, and Cardinal Richelieu, irritated by Saint-Cyran's independence, agreed forcefully: 'Monsieur de Saint-Cyran is more dangerous than six armies.'[18] In 1638 Richelieu had Saint-Cyran placed in Vincennes prison and all his papers seized; nevertheless it was there that Cyran read Jansen's tract the *Augustinus* (published in 1640), outlining Jansenism's main principle, that of St Augustine of Hippo's doctrine of grace. Cornelius Jansen took his place in a long list of commentators all persuaded of the exclusive validity of their understanding

17 The Pascal well at Port-Royal des Champs.

18 Workshop of Philippe de Champaigne, *Jean Duvergier de Hauranne, Abbot of Saint-Cyran*, *c.* 1646, oil on canvas.

of Augustinian thought. To St Augustine is due the following, almost hedonistic, phrase: 'We have to act in conformity with what pleases us the most.'[19] Jansen interpreted this as meaning that the wish of the human is necessarily submitted to the will of the moment and is thus impotent for virtue or vice. Only efficacious grace can turn him to celestial delight. That efficaciousness is not necessarily granted, being rather a gratuitous gift from God.[20] So the Jesuits accused Saint-Cyran of 'recooked Calvinism', and he was brought to trial under Jean de Laubardemont – the judge who ruled over the case of the 'possessed' of Loudun – but was freed on 6 February 1643, after the death of Richelieu, and died the following October. Cardinal Richelieu had declared that Jansen's writing was a plot against the Catholic Church, and Cardinal Mazarin, succeeding him after his death, agreed, trying to push through the papal bull in 1643 that condemned the *Augustinus* for the heretical nature of its five famous propositions.

Antoine Arnauld the father, famous lawyer and brother of both Mère Angélique and Mère Agnès, as well as a close friend of Henry IV of France, had pleaded in 1594 for the Université de Paris to obtain the exile of the Jesuits (calling them thieves and assassins) because of their attacks against the king.[21] Ten of his twenty children were attached to Port-Royal, and even his widow in 1619 entered Port-Royal des Champs. Sainte-Beuve described Arnauld,

> in all his rigor, in all his veracity and his certainty, a right and sure soul doctor . . . *To cure, to cure,* is his only motto, his only care and his only cry: how few limit

> themselves to those! *To wash, to purge* anything that distresses every soul and *defames it before God* . . . The human soul, individual, each soul one by one, naturally and incurably ill through sin, this soul to save by Jesus Christ and only by him, that is his work; he concentrates on it; to the right, to the left, nothing.[22]

In Paris, Arnauld met Cornelius Jansen when he was still a theology student and studied Augustine with him. He was a man of severe asceticism but full of ecstatic optimism, a defender of Gallicanism and of the idea of secrecy, as exemplified in Mère Agnès' *Le Chapelet secret*. (It seems to me that Pascal's own secrecy about his testimonial sewn inside his clothes and never revealed until his death follows in this tradition.) Jean Racine wrote a poem-portrait about Antoine Arnauld the son, the youngest of his father's twenty children, and called 'Le Grand Arnauld' to distinguish him from his father. A famous poem concerning him is entitled simply 'Pour le portrait de M Arnauld'.[23]

To Mazarin's papal bull, as well as to the following affair, therefore, Antoine Arnauld was provoked to respond. A Jesuit priest, Père de Sesmaisons, had authorized his penitent (the Marquise de Sablé) to go to a ball after receiving communion, and admitted sinners to commune immediately after confession. He maintained that the more lacking you were in grace, the more you should dare to take communion. Arnauld refuted this claim on 25 August 1643 in *De la fréquente communion*, which was able to reach a wide range of readers since it was written in French and not the usual Latin. Around the same time, a vicar of Saint-Sulpice refused absolution to the Duc de Liancourt because of his attachment to the Jansenists and to Port-Royal, and so in relation to this condemnation Arnauld wrote again; but he was censured and expelled from the Sorbonne on 31 January 1656. It was subsequently in response to this controversy and the drawn-out battle between the Jesuits – close to Rome – and the Jansenists that Arnauld asked Pascal to write a persuasive text, claiming that he should do something, being young enough to do so. And so he did, writing his *Lettres provinciales*, the subject of a future chapter.

JESUITS AND JANSENISTS

The quarrel between the Jesuits and the Jansenists over salvation was of course of primary importance in Pascal's life and work. The Jesuits were more closely tied to the pope, and the Jansenists to the crown. Françoise Hildesheimer in her authoritative book on Jansenism makes the definition clear: 'What is improperly called Jansenism is simply a reaction against the impious theories of those who have exalted free will to the detriment of divine power; it is a proclamation of the rights of God opposed to an audacious declaration of the rights of man.'[24] Here she is quoting Augustin Gazier, and she continues, describing 'the controversy which has risen in all the epochs of the life of Christianity: the thorny question of the relations between man and divinity which, in the seventeenth century was raging with the Augustinian studies in Louvain, [and] in Holland.'[25] In their turn, the Jansenists condemned Jesuitism as too influenced by Pelagianism.

At the end of the fourth century, the monk Pelagius came to Rome from England, where he emphasized human will, diminishing the action of divine grace. After 410, Pelagius spread his doctrine in Egypt, Palestine and Africa, and was condemned by the popes Innocent I and Zozimus, and by the Council of Ephesus in its last session. In 418 that concilium defined the effects of original sin and affirmed the necessity of divine grace.[26] St Augustine intervened in this context, and laid the basis of a theory of grace to which successive centuries would constantly refer, 'making of God the author of human will, but adding that it is man who makes this will valid; they tend to prefer what comes from man, free will, to what comes from God, will itself'.[27] He proposes the philosophy of Plato and the psychological analogy as most apt to translate the evangelical message: 'Placing in God the foundation of the Ideas which find their efficacious reality in him, he deduces from that an explanation of the order and the contingency of the world . . . He states firmly his position that God is the absolute master of salvation, his grace inspiring desire itself.'[28] For him, no man is good and virtuous without the gift of God the creator, which is grace because it is gratuitous.

After perhaps appearing to be part of the Reformation, after the troubles of the religious wars, France remained Catholic for the most part. After 1627, Protestantism certainly had a lesser impact, as Catholic Reformation took hold and the monarchy strengthened, thanks in large part to Henry IV. In the seventeenth century, nevertheless, reform of the old religious orders was attempted, Port-Royal being an eminent example. Augustinianism dominated the French Counter-Reformation, with the French Church and monarchy on one side, while the adversaries in the Catholic

Counter-Reformation were allied with the *ultramontane* – that is, beyond the mountains over in Italy, the pope. The Duchesse d'Orléans, known as Madame, spoke out for 'Gallican freedoms', alleging that in France, 'we don't care about Rome or the pope!!!'[29] Thus the French Church was specific and autonomous in the Catholic Church, and the lively oppositions between the king and the pope, between two absolutisms, increased.[30]

The Jesuits remained known for a highly intellectual relation to the world and everything else. Before the Council of Trent, 1545–63, and after the spiritual journey of Ignatius Loyola, the Compagnie de Jésus was characterized mainly by obedience to the pope.[31] Let me make a personal statement about what it is to encounter the spirit of St Ignatius Loyola as a young person, for the first time. The exercises of St Ignatius are remarkably visual and hypnotizing, and can have a lifelong effect. I can still see Mary, heavy, coming down the path on her steed, looking for a place to stay: you are meant to feel along with the acts of the drama you place before yourself.

So I come to this project both interested in the Loyola tradition of the Jesuits, and having studied at the Institut Catholique in Paris, which although Benedictine in spirit has a firm tradition of openness to both sides of the Jansenist/Jesuit controversy. And subsequently, through studying Roland Barthes, and his brilliant *Sade, Fourier, Loyola*, I have felt over the years still more drawn to the tensions of the controversy, believing that tensions give life to thought.

PASCAL'S PORT-ROYAL

In 1643, after the death of Saint-Cyran, Antoine Singlin took over Port-Royal de Paris (illus. 19, 20), which was renamed Port-Royal du Saint-Sacrement in 1647. But in 1637 in the abbey in the valley of Chevreuse, where the Solitaires made their retreat on a neighbouring farm called Les Granges, Saint-Cyran had started the *petites écoles*, and until 1660 daily instruction was given there by the bachelors, who had renounced worldly things. It has been suggested that this exercise was intentionally to make that abbey, with its construction based on the swamps, appear a bit more salubrious than it actually was. Having not much of an institutional structure, it was all the same (or perhaps for that very reason) an excellent education for the young. Unsurprisingly, jealousy was aroused, particularly on the part of the Jesuits. The great dramatist Jean Racine, who studied there between 1655 and 1658, proclaimed: 'There has never been an asylum where innocence and purity were further from the contagious air of

19 Port-Royal, engraving, 17th century.

the century, nor any school where the truths of Christianity were more solidly taught.'[32] Pascal had instituted at Port-Royal a new method of learning how to read, based on the idea of discussion instead of memorization, drawing his inspiration from Montaigne's 'De l'art de conférer'. He was vitally interested in education, and in fact educated his two older nephews, Étienne and Louis Périer.[33]

In any case, the nuns returned to Port-Royal des Champs in 1648, and both houses of Port-Royal would exist until 1667. The intention of Port-Royal des Champs was to practise 'total equality, manual work and the life of the spirit away from the epoch but not in total separation from the world'.[34] On the enforced closing of the *petites écoles* in 1660, when Pascal was working on the *Pensées*, Racine asserted that it was 'this instruction of young people that was one of the main reasons which led the Jesuits to the destruction of Port-Royal'.[35] After the death of Cardinal Richelieu, and then Louis XIII, the spirit of Port-Royal, according to Sainte-Beuve's majestic book on the subject, seemed extinct, no longer living except for a few remaining traces in the Jansenism that followed.

But Port-Royal's impact on the Pascal family cannot be overstated, as Blaise lost his sister to its walls, and perhaps part of his own sense of identity once she had gone. He too sought the solitude it represented and the spiritual promise that relinquishing the world seemed to offer. Perhaps he was simply inspired by Jacqueline, who wrote her crucial letter of 1652, begging for his approval, from behind its stern and fortified walls:

> If the world expresses some regret about not seeing me any more, rest assured that it is an illusion that would disappear straight away if it were not a question of opposing some material good; since it is impossible that the world should entertain any friendship for someone who isn't part of it, who never wants to be, and who has presently no greater desire than to destroy everything in relation to it by leaving it forever, by a solemn vow and engaging in a life completely opposed to its maxims, and who would give gladly everything she holds most precious to imprint a like feeling in everyone she knows.[36]

Jacqueline Pascal found her place at Port-Royal du Saint-Sacrement as Soeur Sainte-Euphémie, where she became 'director of the convent's celebrated school', devising 'an

20 Port-Royal.

educational program unusual for its theological sophistication [whereby] every pupil was to have a French/Latin psalter as her basic prayer book [and] on feast days, the nun-teacher would deliver her own commentary on the Gospel for the feast'.[37] And when, in 1661, the convent and Jansenism itself were under attack and the French Assembly of the Clergy drew up a formula of faith that condemned Jansen's five propositions from *Augustinus*, challenging the Catholic religion and the question of grace, Jacqueline was not removed from the situation. When 'King Louis XIV ordered all members of the clergy, religious and teachers in his kingdom to sign [the Formulary]',[38] she did as many other Jansenists did at the time, signing under the guise of agreeing to the doctrine of the Catholic Church but not to its declaration being a matter of fact. Essentially, they were prescribing to the articles of faith, but not accepting the charge against Jansen that assumed his five propositions were heresies. As John J. Conley describes, it was Antoine, staunch supporter of Jansen, who conceived of the distinction between *de jure* and *de facto*, matters of doctrine versus matters of fact. The Catholic Church could not enforce matters of fact to which the pope had not agreed. The Formulary was such a matter of fact, since it accused Jansen's *Augustinus* of certain heretical ideas without sufficient support. 'Consequently, Jansenists could sign the condemnation, letting it be known that their signature only assented to the condemnation of the heretical propositions, not to the allegedly erroneous condemnation of Jansenius for having held these positions.'[39]

In a letter to Soeur Angélique de Saint-Jean dated 23 June 1661, Jacqueline expresses her inner turmoil about the

Formulary and that which she and her sisters were being asked to sign: 'I can no longer hide the pain that pierces the bottom of my heart to see that the only people to whom God has entrusted his truth are unfaithful.'[40] She goes on to ask what prevents all the ecclesiastics forced to sign the Formulary from responding with: 'I know the respect I owe the bishops; but my conscience does not allow me to avow that something is in a book which I myself have not found, especially since so many qualified people also confirm that it is not there.'[41] She wonders about the point of the document, and senses the vulnerability of her position, and her sex, but attempts to rise above it with her claim that,

> I know well it is not up to girls [*des filles*] to defend the truth; though we could say, by some sad and inverse encounter of our times, that since the bishops seem to have a girl's courage, girls then must have the courage of a bishop.[42]

As Father Conley writes, Jacqueline ended up signing, under pressure, but indicated her reserve in a postscript. When she died on 4 October 1661, just three months after her fiery letter to Soeur Angélique about their cause and faith, 'she was acclaimed a martyr to conscience [since] she had embodied the right of the nun to be a *théologienne*: to teach, to provide spiritual direction, to acquire a theological culture and in crises of conscience to withhold assent.'[43] Perhaps it is most fitting to end the chapter as we began, with Jacqueline's loving and desperate words to her brother:

It isn't reasonable for me to prefer any longer others to myself, and it is just that they should do themselves a little violence to pay me back for the violence I have done myself for four years. I am awaiting this witness of your friendship above all others, and pray for my engagement which will happen, God helping, on the day of the Holy Trinity.[44]

FOUR

Pascal's 'Worldly Period', 1650–54

APART FROM PASCAL'S ATTACHMENT – a close one – to Jacqueline, it is clear that he had other women pursuing him and that he was involved with them to some extent. If he was celibate, it was not for want of being desirable. He had, in fact, had the opportunity to marry after Jacqueline left him for Port-Royal; as Attali recounts, Charlotte Roannez, the sister of his childhood friend Artus, expressed a fondness for Pascal that entailed a series of peculiar circumstances. Attali claims that Charlotte's attraction to Port-Royal was partially driven by her hope that Pascal might stop her from taking the veil, as he had hoped to do with Jacqueline.[1] Charlotte was only twenty at the time and, having wanted desperately not to marry the Marquis d'Alluye, she rushed to the convent. She would, it seems, have very much loved to be with Pascal, but he, who so adamantly refused for such a long time to let his beloved sister enter the convent, refused now to help his best friend avoid the fate of his sister: to this an extensive correspondence between September 1656 and February 1657 bears witness. Attali depicts an emotional round robin: 'Charlotte loves Blaise, who loves Jacqueline, who loves God. Charlotte pretends to love God to make

Blaise happy. Blaise, without saying it, shoves her into the arms of God.'[2]

Pascal continued to write to Charlotte, answering every letter she wrote to him, but in 1683 Charlotte destroyed all his letters at the deathbed request of the Duc de La Feuillade, whom she finally married. No originals exist, but copies were somehow kept. We see how, in October 1656, Pascal recommended that she take enough time to detach herself from any earthly love, his having, as he says, already left the world two years previously:

> I know that pain has to come to all persons who, converting, destroy the old man themselves: thus, to make room for new heavens. The whole universe has to be destroyed. Certainly you never detach yourself without pain . . . You have to go out after such an upsetting, and woe to those who are going to give birth or are nursing at that time, that is to say, those who have attachments to the world which holds them in it.[3]

His reference to pregnancy seems odd for one who detested the thought of sex even if for procreation, but his call for detachment here resembles his call in the *Pensées* for the quenching of passion in the face of one's belief in God, forgoing such involvements until that behaviour comes naturally. He claims that 'it is false that we are worthy of the love of others; it is unfair that we should want it', and that 'the tendency to the self is the beginning of all disorder.'[4] This need for order haunted Pascal all his life and is reflected in many of the *Pensées*.

In another letter to Charlotte, he suggests that she must 'try not to be afflicted with anything, to take everything that happens for the best'.[5] And again in December the same year he encourages her to remain on course, 'For in the long run, two things are essential to sanctify yourself, pains and pleasure . . . And so, as Tertullian says, you mustn't think that the life of Christians is a life of sadness. You only leave some pleasures for others, still greater.'[6]

And so the strange story goes: pushed into it by Pascal, Charlotte entered the convent in February 1657, despite the opposition of both her mother and her brother, and probably not out of devotion but despair, at nothing but such devotional encouragement from Pascal. However, as in some play in which the heroine is saved by a *deus ex machina* (although it might seem a bit bizarre to invoke the deity in this particular case), all of a sudden, a relative of her mother arrived at Port-Royal with a *lettre de cachet*, requiring the liberation of the 'demoiselle de Roannez' from the convent to her mother! Still in love with Blaise, Charlotte returned to her mother, and later married the Duc de La Feuillade. His insistence that her letters from Pascal be destroyed seems to allude to the nature of their intimacy, the type of closeness the two once shared – and the entire story strikes the reader as a seventeenth-century *Vanity Fair* plot.

PASCAL WAGERING

Pascal's sister Gilberte constructs him as a solitary figure, as the abbey of Port-Royal, the bastion of Jansenism in France, wanted to imagine him – and of course, for a while he was indeed a Port-Royal 'solitaire' – but in point of fact he detested solitude and had numerous friends with whom he conversed and spent time. As an example, his work on probability calculus he undertook at the request of his friend Antoine Gombaud, Chevalier de Méré, and his experiments with the roulette he undertook thanks to Artus Gouffier. Even the apology for Christianity to which he was devoting his *Pensées* was destined primarily for his friend Damien Mitton, an enigmatic personality – 'joueur passionné', as Attali tells us – named three times in the *Pensées*. And the *Provinciales*, those letters of 1656 and 1657, supposed to come from a naive countryperson questioning the Jesuit form of behaviour, were written to defend his friend Antoine Arnauld (illus. 21). The remembrances of Arnauld and Pierre Nicole reduce the stature of this 'frightening genius' in a way itself frightening. Here is Pierre Nicole writing on 2 September 1662, to M de Saint-Calais, about Pascal's death:

> Posterity will not take much notice of him, since what is left of his works won't be enough to make the vast expanse of his mind known: but in truth he isn't losing much: for human reputations and opinions are little . . . Meanwhile, what remains of this great mind are only two or three little works some of which are strictly useless.

History is entitled to its own opinion about these works. In any case, of course, Pascal's real involvement in the world mattered greatly not just to his famous wager, but to his thinking and ways of wanting to persuade. His friendships with the Duc de Roannez, the Chevalier de Méré and Damien

Mitton were essential to Pascal's comprehension of how to address those he would wish to convert: he had to use wit and worldly wisdom. From this and his wager, probability calculus is thought to have sprung. Overall, much worldly knowhow, if not the morality, then the outward disposition, remained with Pascal, despite the family's wishes, and those of Port-Royal.

21 *Antoine Arnauld, Doctor of the Sorbonne*, 17th-century engraving.

The ideals of good conversation and good company were never explicitly rejected by him. In the worldly spirit, the objective was to partake of the 'bon air' – on the same level as the *bel esprit*, who, according to the 1694 dictionary of the Académie Française, intended for the 'honnête homme', is thought of as having a good spirit, being prepared for good conversation and being always good company.[7] Though Pascal would claim that only the Christian religion 'makes man altogether lovable and happy', since 'in polite society' we cannot be both,[8] the aphorisms of the *Pensées* are perfectly suited to this spirit and this kind of conversation among cultured beings. A certain kind of care for what was useful to persons never left this ultra-religious thinker, so that, as we know, in the very last year of his life, he and Artus Gouffier set up a public transport system that has endured, in the same way that his arithmetic machine has endured; though with changes, to be sure. But the original impulse of these and so many other things was that of the complex and mystic Blaise Pascal, not as Montaigne would have pictured him, always gloomy.

Tales of Pascal out in the world abound, recounted in the Le Guern edition of the Pléiade (1998–2000), the second one after that of Jacques Chevalier in 1954, and also in Nicholas Hammond's *Cambridge Companion to Pascal*, among which is a delightful anecdote about numbers befitting a mathematician. Apparently, Pascal had gone to pay a visit to Arnauld with his long-time friend Gouffier. While strolling around the countryside of Saint-Victoire, the three of them came across a herd of sheep, and Gouffier asked Arnauld if he could guess the number of sheep, of which he claimed he could not. But Pascal, counting them on his fingers in an instant, said there were four hundred. When Gouffier asked the shepherd how many there were, he confirmed the number was in fact four hundred.

THE WAGER

> Let us weigh the gain and the loss in betting that God exists. Let us assess the two cases. If you win, you win everything; if you lose, you lose nothing. Wager, then, without hesitation that he exists![9]

This is Blaise Pascal's famous wager, from his *Pensées*, or 'thoughts', those fragments originally conceived as an apology for the Christian religion. I have not wanted to add to the many books about 'Pascal's Wager', a topic which falls into the category of cliché, since this is the first thing many hear about Pascal, thus overlooking the subtlety of his character and his work. This seems to me atrociously limiting to the shape of his passionate being and even his wit. Laurent Thirouin explains how the wager – that is, the game or gamble – fits in with Pascal's intense efforts at persuasion, relating to the Christian religion:

> The perfect Christian plays a game, this is the human lot, and he accepts the gamble, through the spirit of humility . . . the player acts, but without giving himself away entirely: he somehow always stays, in some fashion, outside his act . . . The *Pensées* are an apology that Providence has transformed into a game.

As Pierre Humbert puts it succinctly: 'Pascal renewed geometry, upset physics, founded probability theory, and applied new and fecund methods to the art of the game.'[10]

Jon Elster's delightfully understandable piece on Pascal and decision theory shows how 'in a caricatural form the key elements of the Wager were already present in the Jesuit writings', ones Pascal challenged in his *Lettres provinciales*, particularly their ideas about God's 'minimal and undemanding' requirements for salvation.[11] Elster suggests that similarities between the two sides are worth remarking on, since 'the Jesuit argument is that a small secular sacrifice will ensure salvation with certainty [whereas] the Wager argument is that a large secular sacrifice will ensure salvation with some non-zero probability.'[12] The expected payoff in both cases, as Elster says, is an infinite one. Here Pascal merges religious belief with scientific and mathematical culture.

As a mathematician, and a theoretician of probabilities, mathematical constructs are of great importance for Pascal. He claims that 'we know the existence of the infinite and do not know its nature, because it has extension like us, but not limits like us. [Whereas] we do not know either the existence or the nature of God, because he has neither extension nor limits.'[13] Reason cannot determine the question of God, but it is rational to make some attempt at it, thus setting the question in play. He claims that 'at the extremity of this infinite distance, a game is being played in which heads or tails will turn up.'[14] It is worth remarking that the French word he uses for 'heads' is *croix*, which is translated as 'face', referring to the heads portion of a coin, but of course *une croix* is also 'a cross', so that the symbol enters into the wager.

Now of course, here as elsewhere, translation matters enormously. I shall give one or two examples, starting with the world-famous conclusion of the world-famous wager, in the 'développements de 1658–1662', reading in the original:

> Le juste est de ne point parier.[15]
> Oui, mais il faut parier. Cela n'est pas volontaire, vous êtes embarqué.

Forget the number of times we read something like, 'you are embarked'. Although, of course, it is good to bring back the water imagery from another well-known passage from the 'Disproportion de l'homme': 'Nous voguons sur un milieu vaste, toujours incertains et flottants, poussés d'un bout vers l'autre' (Disproportion of man: We are wandering over a vast surface, always uncertain and floating about, pushed from one end towards the other).[16] Moreover, as some commentators remind us, this watery imagery – enhancing the flow of the text as it carries us along – that watery basis for our sliding and slipping, our constant search for a stable base ('Rien ne s'arrête pour nous', despite his other assertion that movement is our life, repose is our death) may spring from Pascal's early upbringing in the seaports of Rouen. And of course we hear, from Montaigne's *Essais*, what Pascal read therein: '[La raison] se perd, s'embarrasse et s'entrave, tournoyant et flottant dans cette mer vaste, trouble et ondoyantes des opinions humaines, sans bride et sans but' ([Reason] is lost, stumbles and gets entangled, turning and floating in this disturbed and wavering sea of human opinons, with neither rein nor purpose.)[17]

So we pay particular attention to these set pieces, about our inconstancy and our necessity of decision-making. They are in fact the passages that still now and always continue to seize us anew. Here is Roger Ariew's version, so far from the flatly literal 'you are embarked', which rouses no little hairs on the back of any neck:

> The right thing is not to wager at all.
>
> Yes, but you must wager. It is not optional. You are committed.[18]

I feel right away that, indeed, I am committed. It is very like the debates about the best translation of the existential term of 'engagement', for which we often use the term 'involved'. I would be happy, in general, to use this 'committed' . . .

In a letter to Pierre de Fermat, a mathematician from the Mersenne circle with whom he discussed probabilities in depth, he claimed that his friend the Chevalier de Méré (Gombaud) was unable to understand that a mathematical line is infinitely divisible,[19] which to Pascal is as simple as accepting that God exists. It was in fact for Méré – a gambler – that he had begun his work on discovering how to measure, essentially, the risk of a decision; about the 'points' involved in a game and who wins if the game is interrupted. In a different fragment in his *Pensées*, he discusses diversion as a method for managing a life, suggesting that

> a man lives his life free from boredom, playing every day for a small amount. Give him every morning the money he can win that day, on the condition that he

> does not play: you make him unhappy . . . It is not, therefore, only the amusement he seeks . . . He must fashion for himself an object of passion to excite his desire, his anger, his fear for the fashioned object.[20]

This fragment was the most worked over by him, and this discussion of ennui is particularly fresh for the time. The game, famously known as Pascal's Wager, is referred to widely in cases requiring a risk with the potential for infinite gains against finite losses. The gamble itself is necessary, as the universe has been set in play, and not to wager is unacceptable:

> Yes; but you must wager. It is not optional. You are committed . . . You have two things to lose, the true and the good; and two things to stake, your reason and your will, your knowledge and your beatitude; and your nature has two things to avoid, error and wretchedness. Since you must necessarily choose, your reason is no more offended by choosing one rather than the other. This settles one point. But your beatitude? Let us weigh the gain and the loss in calling heads that God exists [*croix que Dieu est*]. Let us assess the two cases. If you win, you win everything; if you lose, you lose nothing. Wager, then, without hesitation that he exists![21]

Much is included, predicted and relevant here. The pronoun 'you' involves us not as bystanders but as participants, and it is easy to see how some present-day stock market analysts refer to Pascal to explain their own risk-taking. Pascal refused at the end of his life to indulge what he saw as the triviality of the argument between Jesuits and Jansenists: that was not part of the game.

THE GEOMETRY OF CHANCE: PASCAL AND FERMAT

As mentioned earlier, Pascal corresponded with Pierre de Fermat, a genius at geometry who never actually published anything in his lifetime but shared his ideas openly through conversations and in letters, as well as in marginal notes written into the texts of others. He was a lawyer for the Toulouse parliament but is credited with discovering early components of differential calculus, contributing much to number theory and analytical geometry (illus. 22). Pascal met Fermat when he was still a young boy, visiting Mersenne's circle with his father, listening to him discuss geometry with Descartes, but began to correspond with him only once he started working on mathematical probabilities, or analysis of mathematical outcomes, more than twenty years later. In a letter dated August 1660, six years after their original exchange and two years before Pascal died, he addressed Fermat as the most 'gallant' man in the world and assured him he was capable of discerning such qualities better than anyone since he admired them infinitely, especially when tied to someone as talented as Fermat. He claimed that if he were in better health he would visit Fermat in Toulouse, not because of his being the greatest geometer in Europe but because he imagined an abundance of 'esprit' and 'honnêteté' in their conversations.[22] To speak plainly about geometry, Pascal wrote, is the highest exercise of the mind, though he admitted it was a 'pointless'

endeavour; he considered it 'the best occupation in the world [but] still only an occupation'.[23]

The 1654 correspondence between the two mathematicians, essentially denoting an early theory of probability, seems to make up one tract, offering examples and showing their reliance on each other's work as they move forward in their search for an equation that satisfies the original question, of measuring an outcome for a game that remains unfinished. Their letters emit a generosity that befits Père Mersenne's

22 Francois de Poilly, *Pierre de Fermat*, 17th century, engraving.

community of shared ideas and intellectual rigour, but they are also polite and complimentary without a hint of ego or competition. In a letter to Pierre de Carcavi, in fact, Fermat suggests that the work, should it be published, should appear without his name, and that Pascal should be considered the author.[24] His admiration and respect for the younger mathematician, who had so brilliantly contributed to the field of projective geometry with his treatise on conics, are evident in this letter, as he opens it by expressing his pleasure at having found Pascal's feelings about the project to be in line with his own. Also, in an effusive manner quite typical of the age, he 'greatly esteems [Pascal's] genius and believes that he is capable of conquering everything he undertakes'.[25]

Pascal would perhaps disagree, as he does in his October letter in response to Fermat's request that he pursue his mathematical proposition that 'there is no such triangle in whole numbers whose area is equal to a square number.'[26] Fermat's contribution here comes from a marginal note he made to a missing proof in the work of Diophantus, an Alexandrian Greek mathematician who was called 'the father of algebra'. Fermat assumes not only that Pascal will understand his proposition, but that through it he will make innumerable new discoveries, claiming 'multi pertranseant ut augeatur scientia' (may many surpass so that knowledge be enriched).[27] In response, Pascal first expresses his admiration for the very different method of chance detailed in the letter, as it confirms that their methods are distinct and individual while equally elegant. But with regard to Fermat's 'numerical inventions for which he has been gracious enough to send the specifics',[28] Pascal confesses that they are beyond him and that he may only admire them, humbly beseeching Fermat to complete them at his leisure. Le Guern suggests that this intellectual humility and refusal may not, in fact, refer to the proposition stated above, but rather to his taking up Fermat's earlier proposition in the letter dated 29 August, which suggested that Fermat had made an error in his calculations. Wanting to spare him any embarrassment, Pascal refused to take it up, assuming Fermat would eventually catch the mistake himself.[29] Either way, Pascal's modesty and reverence for the geometer are plain, and through their correspondence we are privy to how genius operates, flourishing when corresponding with its own kind.

In Pascal's particular genius, especially with regard to his scientific and mathematical rigour, we see his need for hard proof to support his findings, whether numerical or physical in nature. All the more significant, then, is the fragment known simply as Pascal's Wager, in demonstrating to what extent Pascal is attuned to human psychology. Though he defends a Christian's right not to offer proof of God, or evidence of his nature, he treats the matter as though God were in fact a natural law, or an observable and inevitable phenomenon. Proof of God is not required for faith. We might invoke Tertullian here: 'prorsus credibile est, quia ineptum est' (it is by all means to be believed, because it is absurd).[30]

The laws of logic dictate that if one were given evidence that God exists, one would no longer need to act by faith alone, which is essentially the point. In a 1656 letter to Charlotte, Pascal insists that 'if God were to continually reveal himself to man, there would be no merit in believing in him.'[31] Faith for him is, if I may put it in light fashion, something we have to work at and put into play in order to deserve it.

FIVE

The 'Second Conversion': The Memorial; M de Sacy on Epictetus and Montaigne, 1655–7

THE CELEBRATED TESTIMONIAL of the night of fire is usually thought of as Pascal's second conversion, at the age of 31, after the first, at the curing of his father after his fall on the ice, when the thought of the Abbé Saint-Cyran worked its spiritual magic. His second conversion seems to have happened shortly after his own encounter with fate. As Attali recounts, an anonymous tale claims that Pascal was almost in a coach accident when the two front horses of his carriage led it to part of a bridge that had no railing. The horses headed into the water, holding the carriage on the edge of the abyss, though never tipping it over and into the pool. His night of fire, which gave birth to his testimonial, was inspired by a conversation with Antoine Singlin at Port-Royal, followed by a two-day retirement alone in a room without food or drink, and only the biblical translations of the Jansenists of Louvain to keep him company.[1] This testimonial, or 'Memorial', which Pascal sewed into whatever he was wearing, inscribed on a bit of parchment and enfolded into another folded pouch, was found

only after his death and was something he never discussed but lived by and kept hidden in his clothes, just as he also wore a hair shirt. It haunted his memory and has haunted many of ours over the years, this fiery testament to his belief, which he kept close to his skin and never spoke of. It is the secrecy of this so personal act and its endurance, along with its drama ('Joy, joy, tears of joy'), that solicit our attention.[2] My first engagement, very many years ago, with this very odd genius began with my reading of this text, which I give in full and in translation:

23 Philippe de Champaigne, *Antoine Singlin*, *c.* 1664, oil on canvas.

L'an de grâce 1654,
Lundi, 23 novembre, jour de saint Clément, pape
et martyr, et autres au martyrologe. Veille de saint
Chrysogone, martyr, et autres,
Depuis environ dix heures et demie du soir jusques
environ minuit et demi,
FEU.
«DIEU d'Abraham, DIEU d'Isaac, DIEU de Jacob»
non des philosophes et des savants. Certitude.
Certitude. Sentiment. Joie. Paix. DIEU de Jésus-
Christ. *Deum meum et Deum vestrum.* «Ton DIEU sera
mon Dieu.» Oubli du monde et de tout, hormis
DIEU. Il ne se trouve que par les voies enseignées
dans l'Évangile. Grandeur de l'âme humaine. «Père
juste, le monde ne t'a point connu, mais je t'ai
connu.» Joie, joie, joie, pleurs de joie. Je m'en suis
séparé: *Dereliquerunt me fontem aquae vivae.* «Mon Dieu,
me quitterez-vous?» Que je n'en sois pas séparé
éternellement. «Cette est la vie éternelle, qu'ils
te connaissent seul vrai Dieu, et celui que tu as
envoyé, Jésus-Christ.» Jésus-Christ. Jésus-Christ.
Je m'en suis séparé; je l'ai fui, renoncé, crucifié. Que
je n'en sois jamais séparé. Il ne se conserve que par
les voies enseignées dans l'Évangile: Renonciation
totale et douce. Soumission totale à Jésus-Christ
et à mon directeur. Éternellement en joie pour un
jour d'exercice sur la terre. *Non obliviscar sermones tuos.*
Amen.

In English translation:

The year of grace 1654,
Monday, 23 November, feast of St Clement, pope
and martyr, and others in the martyrology.
Vigil of St Chrysogonus, martyr, and others.
From about half past ten at night until about half
past midnight,
FIRE.
GOD of Abraham, GOD of Isaac, GOD of Jacob not
of the philosophers and of the learned. Certitude.
Certitude. Feeling. Joy. Peace. GOD of Jesus Christ.
My God and your God. Your GOD will be my God.
Forgetfulness of the world and of everything, except
GOD. He is only found by the ways taught in the
Gospel. Grandeur of the human soul. Righteous
Father, the world has not known you, but I have
known you. Joy, joy, joy, tears of joy. I have departed
from him: They have forsaken me, the fount of living water. My God, will you leave me? Let me not
be separated from him forever. This is eternal life,
that they know you, the one true God, and the one
that you sent, Jesus Christ. Jesus Christ. Jesus Christ.
I left him; I fled him, renounced, crucified. Let me
never be separated from him. He is only kept securely
by the ways taught in the Gospel: Renunciation, total
and sweet. Complete submission to Jesus Christ and
to my director. Eternally in joy for a day's exercise on
the earth. May I not forget your words. Amen.[3]

Privacy and memory: how important these now seem and indeed feel to us. Despite his so obvious illnesses, the convulsions, the crutches, the almost incessant pain and the fact that he had to abandon his scientific research and indeed, finally, his writing, Pascal never betrayed his own testament, which he kept hidden and treasured. This fact seems to me crucial for our understanding and appreciation of his overwhelming sincerity and utter humility, even about what was the closest to him.

The various pseudonyms under which Pascal wrote; the ability to persuade his worldly friends – even the ones he did not convert to his way of thinking; his very ardent defence of his friends, like Antoine Arnauld; his final acceptance of his sister's vocation – all of this persuades me that his rapid and easy abandonment of things he too readily understood were part of his nature and a large part of the reason we are not wrong to dwell on his character beyond his inventions and genius. I am not converted, but I am persuaded, far beyond his 'art of persuasion'. Everything contributes to my persuasion, and I see no reason to linger longer on this most remarkable of testaments, kept silently, secretly, and all the more extraordinary for it.

THE *ENTRETIEN* WITH M DE SACY

Among the important debates at Port-Royal, Pascal's discussion with Le Maître de Sacy (Louis-Isaac Lemaistre de Sacy), the director of the abbey, about philosophy, as recorded by Sacy's secretary, Robert de Fontaine, in the year 1654 or 1655, remains prominent.[4] I will report on it as Fontaine passed it down, as it is extracted from his *Mémoires*. He begins by describing the spirit of peace encouraged by the Maître de Sacy, who insisted that we should never judge, 'because only God can judge with certainty',[5] and how he would calm all conflicts that would arise in the sciences of theology and ordinary natural disputes. 'How many little agitations there arose in this desert', writes Fontaine about Port-Royal des Champs, 'in regard to the human sciences of philosophy and the new opinions of M Descartes!'[6] In the chateau of the Duc de Luynes, a friend of Port-Royal, everyone talked all the time about Descartes' new system of the world, or Cartesian thought, and its three separate areas of existence: the corporeal, the mind and God. Fontaine claims that de Sacy 'admired the conduct of God in these new opinions of Descartes, where everyone bowed their head', and simply accepted 'Descartes and Aristotle as a thief who came to kill another thief and take his spoils'.[7] It bothered Sacy that the work of these philosophers was treated as equal to Scripture. That everyone's ideas were killing those of another was great, he said: 'All the better. More dead, fewer enemies. Maybe the same will happen to M Descartes.'[8]

Sacy claimed that God made the world for two reasons: the first was to show his grandeur, and the second to paint invisible things within nature. But, he believed, Descartes' ideas destroyed both. He accuses him of ignoring the whole, for the sake of admiring and questioning each part or material in its composition. Fontaine quotes what Sacy may have imagined Descartes saying in response:

> I'm not claiming to tell things as they really are. The world is such a big object that you get lost in it; but I look at it like a number. Some turn and turn round again the letters of this alphabet, and find something;

> me too I've found something, but it isn't perhaps what God made.[9]

Sacy says that some grope for the truth, and so if they find it, it's just luck. He compares it to the clock over the bridge of Notre-Dame: 'The clock was telling the truth then and I said: Let's go quickly, it won't be right soon. It's the truth now, but it only tells the truth once a day.'[10] Fontaine hoped not to be dazzled by the brilliance of Pascal, who was in any case charming and persuasive. Sacy realized early on that 'everything Pascal said, he himself had already read in Saint-Augustine',[11] and so he credited Pascal with having discovered the same truths as those of the early Church Fathers. As it was Sacy's habit to speak with those about what they knew, he happily talked with Pascal about philosophers whom he had not read widely himself. For example, they discussed Epictetus, a stoic philosopher who believed we have no control over external events since they are determined by fate, but that we may control our actions with rigorous self-discipline. In his interview with Pascal, as Fontaine recounts, Sacy claimed that Epictetus dealt with loss as a true philosopher:

> 'Never say', he said, 'I lost that'; say rather 'I gave it back.' My son died? 'I gave him back.' My wife died? 'I gave her back.' Same thing about property and everything else . . . You shouldn't wish things that happen would do so as you would wish; but you should wish they happen as they happen. Remember, he said, 'that you are here like an actor, and you play the role

> of a play just as it pleases the master to give it to you. If he gives it to you short, play it short; if he gives it to you long, play it long.' 'It's up to you to play the role you are given; but choosing it, that's up to someone else. Contemplate death every day, and don't want anything with excess.'[12]

He also claimed that the philosopher showed a thousand ways man must behave, never boasting about his accomplishments, but learning God's will and following it. Pascal, however, suggests the ways in which Epictetus had got it wrong, not recognizing his powerlessness in the sight of God:

> the spirit cannot be forced to believe that which it knows to be false, nor can the will love that which it knows may make it unhappy; that both these powers are therefore free and it is by them that we may become perfected; that man can, by these powers, know God perfectly, love him, obey him, please him, be cured of all vices, acquire all virtues, become a saint, friend and companion to God.[13]

For Pascal, these errors led Epictetus to make others that were even graver, such as the idea that man may take his own life if he thinks God calls him to do so.

They also discussed Michel de Montaigne, the influential sixteenth-century French philosopher most famous for his *Essais*, those tryings-out of ideas. For Montaigne, as Pascal had said,

> puts everything in a universal doubt, so general that this doubt gets carried away with itself, that to say he doubts if he doubts, and doubts even this last proposition. His uncertainty turns about itself in a perpetual circle without resting . . . It is in this doubt that doubts itself, and in this ignorance which ignores itself; and what he calls his master form, which is the essence of his opinion that he could not express in positive terms . . . not wanting to say: 'I don't know', he says 'What do I know?' which he makes his motto . . . He destroys everything that passes for the most certain thing among men . . . to show that you don't know where to settle your belief.[14]

Montaigne, 'being situated with such an advantage in this universal doubt . . . is fortified equally by its triumph and its defeat'.[15] Pascal believes it is in this wavering, bobbing and weaving that Montaigne, the professed Catholic born in a Christian state, can combat and claim victory over the heresy of his sixteenth-century France. 'Montaigne is incomparable to confound the pride of those who without faith think they know true justice; to disabuse those who are attached to their opinions, and who think they find unshakeable truth in sciences.'[16]

According to Fontaine, Sacy, listening avidly to Pascal, invoked these words of St Augustine in praise of the things he had been shown: 'Oh God of truth! Are those who know these subtleties of reasoning more agreeable to you?'[17]

PASCAL AND DESCARTES, AND MONTAIGNE IN THE MIDDLE

As for the relations between Pascal and Descartes, there is much to be said, of course, since he and Pascal were not at all congenial, to understate the case.[18] In her masterful *How to Live*, Sarah Bakewell discusses, with brevity and wit (two things Pascal had in spades), the whole nest of complications around Descartes, Pascal and, of course, Montaigne. No surprise, given the title, that the latter comes out as infinitely more approachable and moderate than the other two. As T. S. Eliot said of Montaigne, he was for Pascal 'the great adversary' – because he would win every time. As Bakewell puts it: 'He found the Pyrrhonian tradition, as transmitted through Montaigne, so disturbing that he could hardly get through a page of the *Apology* without racing to his notebook to pour out violent thoughts about it.'[19] He would continually circle certain thoughts of Montaigne, consider them, reacting to such a degree that their dialogue is irreplaceable. The Pyrrhonian tradition relies on scepticism as its base for inquiry, considering absolute certainty of knowledge impossible. However, a Stoic such as Epictetus, calm in every case, is essentially impervious to misfortune.

Here I feel it necessary to stress only a few things that strike me as relevant. First, Montaigne's memory of his riding accident – when he was knocked senseless from his horse, unintentionally, by a horseman behind him – and that of his recovery stayed with him always, just as Pascal's near accident with the carriage on the cliff was never far from his mind. These crucial, no matter how accidental, happenstances marked their thought irreparably with utterly different results. Montaigne remained moderate in his way of living, grateful for his recovery and pleased to write on and on about himself, for the delight of himself and of the ages. Pascal remained haunted by the abyss he sensed on his left side, and this haunting itself was to mark much of what we know of him, mystery and all. As André Breton, the founder of the Surrealist movement, quoted a proverb so memorably at the beginning of his prose poem of a novel, *Nadja*: 'Tell me whom you haunt and I will tell you who you are.' We feel Pascal haunting that abyss as it haunts him. But scarcely could two characters be more opposed in every way, since Montaigne says of himself:

> I have no great experience of these vehement agitations, being of an indolent and sluggish disposition; I like temperate and moderate natures. My excesses do not carry me very far away. There is nothing extreme or strange about them; and the most beautiful lives, to my mind, are those that conform to the common human pattern, with order, but without miracle and without eccentricity.[20]

The 'Apologie de Raimond Sebond', Montaigne's essay and argument for a sceptical Christianity based on faith and not reason, so important for thinker after thinker since Montaigne, has – in its very opposite thesis – the same title and project as Pascal's unfinished *Apologia* for the Christian religion. But Montaigne's 'Apologie' has no feeling of haunting: it does not cling to your imagination as bearing any sort of doom. And as for the Enlightenment figures of Diderot and Voltaire, they appear to us as far from Pascal as does light from dark. Voltaire's grudge against Pascal is so well known that I cannot resist repeating many parts of it, most famously: 'Pascal, you are sick!' and 'I venture to champion humanity against this sublime misanthropist.' And then, after attacking, in his witty way, 57 of Pascal's *Pensées*, Voltaire continues:

> When I look at Paris or London I see no reason for falling into this despair Pascal talks about. I see a city not looking in the least like a desert island . . . To look on the world as a prison cell and all men as criminals is the idea of a fanatic . . . What a delightful design Montaigne had to portray himself without artifice as he did! For he has portrayed human nature itself. And what a paltry project of . . . Pascal, to belittle Montaigne![21]

'Paltry' is a strong word here and, I think, the wrong one, given Pascal's long struggle with exactly the 'human nature' of Montaigne's *Essaís*, so massively important to him that he would circle and re-circle passages in his rereadings. Not paltry, it seems to me; rather, desperately involving of his entire mode of being and believing; less paltry than all-engaging; less belittling than attacking, as best he could. Whichever side we find ourselves on, the words we use are bound to engage our minds. As Gaston Bachelard said, 'the size of our world depends on our vocabulary.' But then, great thinkers and atheists like Voltaire have the right to speak as they wish of great thinkers and believers like Pascal. Montaigne was a great and pragmatic thinker, and Pascal never got over him; nor indeed do we think that he should have done so.

SIX

The *Provinciales* and the Miracle of the Holy Thorn

THE ONGOING STRUGGLE between 'the two modalities' – that is, sufficient and efficacious grace – continued with a vengeance. Pascal, under the name of Louis de Montalte (no one knew his identity until Pascal was dead), wrote his first letter on 23 January 1656.

An anagram of Louis de Montalte, the very name of Salomon de Tultie echoes a verse of the First Epistle to the Corinthians: 'Because the world has not known God in wisdom through wisdom, it pleased God through the madness of the one who preaches [*per stultitiam praedicationis*] to save believers.' Which Pascal resumes: 'Contraries. Infinite wisdom and madness of religion.' The name Salomon indicates, from ancient times, a wisdom, and under Tultie we recognize *stultitia*, folly or foolishness, thus forming again a convergence of contraries. Salomon de Tultie's papers are mixed with Pascal's and often cannot be distinguished from them. Salomon's style is more usual, insinuates itself more easily, sticking in the memory, and is more often quoted because it is about ordinary things of life. Like Jesus Christ, St Paul and St Augustine, Salomon chooses 'the order of charity, not of the mind', and 'consists principally of a digression on each point which has some relation to the end, to keep it always in mind', this 'order of charity' being the selfless love of the other, a general but totally authentic, biblically based feeling for fellow humans at large and the love of the singular other in human tenderness.[1]

The first letter was an instant success and was followed by eighteen more 'divine letters', which the Marquise de Sévigné, a French aristocrat and avid letter-writer herself, greatly enjoyed reading. She was enthusiastic about their perfect style, claiming there is no finer mockery or more natural and fastidious a child of Plato's great dialogues, and that the eloquence of the *Provinciales* evinces Montalte's love of both God and truth.[2] The first three letters were linked to the defence of Antoine Arnauld, who had urged Pascal to do something for the Jansenist cause, and the remainder to a counter-attack on the lax morality of the Jesuits. The *Lettres provinciales*, published in a volume in 1657 (that is, *Letters Written by Louis de Montalte to a Provincial Friend and to the Reverend Jesuit Fathers on the Subject of the Morals and Politics of these Fathers*), stated in the first letter that they concern 'the subject of the present debates in the Sorbonne', the dispute over Jansen's *Augustinus* and its five seemingly heretical propositions. After the first eighteen letters, fragments of a nineteenth, addressed to Père Annat, was published in 1779. These documents are a masterpiece of satire and classic prose, used for subsequent centuries to amuse, enliven and serve as a model for satirical and critical pieces.[3]

The *Lettres provinciales* are full of badinage, teasing and real (and efficacious) jabs. The first three are light, but starting with the Fourth Letter, they take on a more serious tone. In the fourth, for example, Montalte as a naive man from the country, and not the city, questions a Jesuit about the term 'grâce actuelle' (actual grace), and through his inquiry provokes an example of the Jesuit's thoughts on redemption. More than happy to instruct the curious, the Jesuit explains that sufficient grace is 'an inspiration from God by which he shows us his will and by which we are excited to fulfil it'.[4] Troubled by the term 'grâce actuelle', Montalte asks for a more thorough explanation. The Jesuit claims that 'an action cannot be considered sinful if God does not make us aware, before we commit it, of its sinfulness, and offer an inspiration that moves us to avoid it altogether'.[5] Cleverly feigning confusion still, Montalte reveals the famously casuistic style of the Jesuit, who proceeds to fetch a stack of books, none of which is an actual Bible, with examples that explain the meaning of the term 'grâce actuelle'.

The first four letters and the last two take up the question of grace and freedom. According to Françoise Hildesheimer, in 1588 the Jesuit Luis de Molina in Lisbon promulgated the theory about the two graces: 'For the "efficacious grace" of the Augustinians, he substitutes a "sufficient grace" which would bring man everything necessary to accomplish good, but would only take effect on his free will.'[6] So God gives humans an adequate quantity of sufficient grace. This is the basis of the argument about sufficient and efficacious grace. Arguing against Molina and his opinion about human freedom, Pascal defends the Augustinian tradition and its determinism. God wants the death of the sin, not the death of the sinner, as he puts it.

The subsequent letters attack the lax morals of the Jesuits, relying on particular cases. From here on, we find very little of the comic or the picturesque of the first letters, but they retain their verve.[7] In 1657, the Parliament of Provence condemned the sixteen first letters; and on 6 September the same

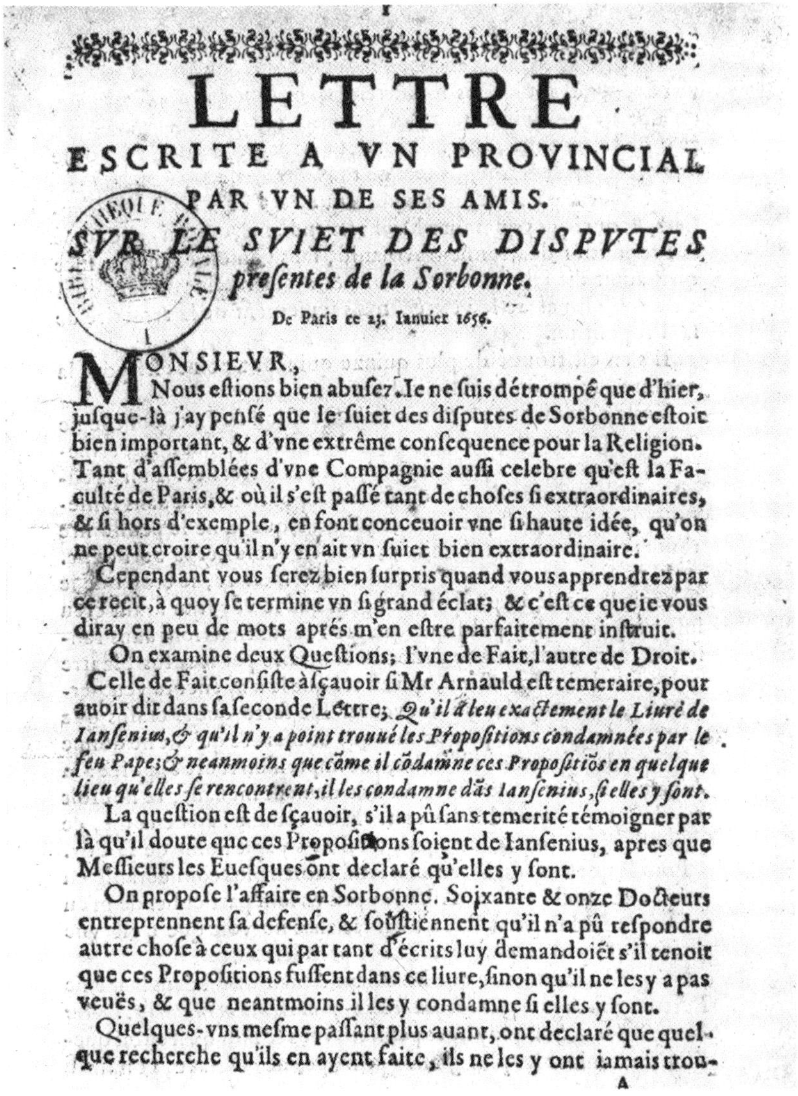

1

LETTRE ESCRITE A VN PROVINCIAL PAR VN DE SES AMIS.

SVR LE SVIET DES DISPVTES *presentes de la Sorbonne.*

De Paris ce 23. Ianuier 1656.

MONSIEVR,

Nous estions bien abusez. Ie ne suis détrompé que d'hier, jusque-là j'ay pensé que le suiet des disputes de Sorbonne estoit bien important, & d'vne extrême consequence pour la Religion. Tant d'assemblées d'vne Compagnie aussi celebre qu'est la Faculté de Paris, & où il s'est passé tant de choses si extraordinaires, & si hors d'exemple, en font conceuoir vne si haute idée, qu'on ne peut croire qu'il n'y en ait vn suiet bien extraordinaire.

Cependant vous serez bien surpris quand vous apprendrez par ce recit, à quoy se termine vn si grand éclat; & c'est ce que ie vous diray en peu de mots aprés m'en estre parfaitement instruit.

On examine deux Questions; l'vne de Fait, l'autre de Droit.

Celle de Fait consiste à sçauoir si Mr Arnauld est temeraire, pour auoir dit dans sa seconde Lettre; *Qu'il a leu exactement le Liure de Iansenius, & qu'il n'y a point trouué les Propositions condamnées par le feu Pape; & neanmoins que come il codamne ces Propositios en quelque lieu qu'elles se rencontrent, il les condamne dãs Iansenius, si elles y sont.*

La question est de sçauoir, s'il a pû sans temerité témoigner par là qu'il doute que ces Propositions soient de Iansenius, apres que Messieurs les Euesques ont declaré qu'elles y sont.

On propose l'affaire en Sorbonne. Soixante & onze Docteurs entreprennent sa defense, & soustiennent qu'il n'a pû respondre autre chose à ceux qui par tant d'écrits luy demandoiẽt s'il tenoit que ces Propositions fussent dans ce liure, sinon qu'il ne les y a pas veuës, & que neantmoins il les y condamne si elles y sont.

Quelques-vns mesme passant plus auant, ont declaré que quelque recherche qu'ils en ayent faite, ils ne les y ont iamais trou-

A

24 Provincial letter of Pascal, January 1656.

year, the pope put them on the index of books forbidden to the faithful. Surely this is the supreme compliment for such writing, speaking to its efficacy, to echo one of the terms involved in the debate. Everywhere in them, there are witticisms that have come down to us, such as the remark in the Sixteenth Letter: 'This letter is longer because I didn't have time to make it shorter.'[8]

Montalte presents himself as neutral, naive and not very well informed. First, he is a fictitious narrator, but later the references are to real people, with a transcription of real dialogue. The figure of Escobar, a real Spanish casuist, well known and whose name is used for that reason, in Pascal's poem on the subject of Port-Royal is taken to represent Jesuit casuistry as a whole. In the Seventeenth Letter Pascal affirms, as the narrator, that he is not of Port-Royal. Pascal found all the continuing debates about grace and predestination to be sterile, tiring and beside the point of real faith. All his efforts could not prevent the successive condemnation or the destruction of Port-Royal. When the nuns had to sign the formulary against Jansenism, Pascal did not have to sign, and of course was radically opposed to it.

Much of the detailed satire in the *Provinciales* is over-the-top amusing. One of my favourite details, among so many in the letters, is about the laxist customs, and it seems to me the height of delicious casuistry. You could go to two halves of two different Masses and it would count as your Sunday duty of attending a complete Mass.[9] The letters seem at first a game, but prove increasingly serious. Excerpts from a few of them display their verve and way of presenting themselves.

> *First Letter*
> It's a matter of examining what M Arnauld said: '*That the grace without which you can't do anything, was lacking to Saint Peter in his fall.*' On that we thought, you and me, it was a matter of examining the greatest problems of grace, if it is given to everyone or if it is efficacious; but we were wrong . . . To get it right, I saw M.N., a doctor from Navarre, who lives near me, who is among the most passionate against the Jansenists; and as my curiosity made me almost as ardent as him, I asked him if they would decide formally that 'grace is given to everyone' so that there wouldn't be doubt any more. But he rebutted me, telling me it was not the point; that there were those on his side who held that grace isn't given to everyone, that the examiners themselves had said right out in the Sorbonne that this opinion is *problematic*, and that he agreed with this himself: which he confirmed to me by this passage, which he said is famous, from St Augustine: '*We know that grace is not given to all men.*'[10]

The First Letter speaks amusingly about the term 'pouvoir prochain', about which there was incessant argument, in this fashion:

> That means, so I say to them, when I'm leaving, that you have to pronounce these words with your lips, for fear of being a heretic by name? For really are these words in the Holy Writ? – No, they say. – So do they come from the Holy Fathers, or from the Concilia or somewhere else? – So what's the need

> to say them, since they themselves don't have any authority nor any meaning? – You are so opinionated! they tell me: either you'll say them or you'll be a heretic and M Arnauld too. For there are more of us; and, if we need to, we'll have so many monks coming that we'll win.[11]

So this is power in numbers, as a threat: what a satire!

Some of the *Provinciales* are very funny indeed: take a passage in the second one that twists itself around the crucial vocabulary of the two kinds of grace, sufficient and efficacious, setting the churchman up for a trap he is bound to fall into. A ruling had just been passed that each speaker had only half an hour to discourse, so the 'naive' interlocutor asks, to begin with:

> *Second Letter*
>
> Do you adjust what you have to say to a special amount of time? Yes, he said, for some days now. And do you have to speak for half an hour? No, you speak as little as you like. But not as much as you like, I said. Oh, what a fitting rule for ignorant people! Oh, the honest pretext for those who have nothing good to say. But really, Father, this grace given to everyone, is it *sufficient*? Yes, he said. And nevertheless it doesn't work without *efficacious grace*? That's true, he said. And everyone has the *sufficient* one, I said, and everyone doesn't have the *efficacious* one? That's true, he said. That means, I said to him, that everyone has enough grace and that everyone doesn't have enough of it; that means

> that this grace suffices, although it doesn't suffice; that means that it is sufficient in name and insufficient in fact. Truly, my Father, this doctrine is very subtle. Have you forgotten, in leaving the world, what the word *sufficient* means out there? Don't you remember that it encloses everything that is necessary too? How can you say that everyone has a grace sufficient to act, because you admit that there's another absolutely necessary in order to act that everyone doesn't have? Is this belief of no importance and are you leaving it up to people's freedom whether efficacious grace is necessary or not? Is it indifferent to say that with sufficient grace people can act in fact? What, said this gentleman, indifferent! It's a heresy, it's a formal heresy. The necessity of efficacious grace to act effectively is an article of faith; it's a heresy to deny it.[12]

Montalte, as the naive interlocutor, continues wondering: if he denies sufficient grace, does that alone make him a Jansenist? He claims, 'If I admit it like the Jesuits so that efficacious grace isn't necessary, I will be a heretic . . . And if I admit it like you, so that efficacious grace is necessary, I am sinning again common sense, and I am outlandish, say the Jesuits.'[13] So, he says, he must either be considered outlandish, or heretical, or a Jansenist. Delightful.

Here is a stand-alone treasure, being the 1657 version of the Sixth Letter:

> You see that, whether through the interpretation of terms, or by noting the favourable circumstances, or

> the double probability of the for and the against, we always arrange these so-called contradictions that astonished you before, without ever hurting the decisions of the Sacred Writings, the Concilia or the Popes, as you see. My Reverend Father, I said, how the Church is lucky to have you as a defender. How useful are these probabilities![14]

How lucky is the reader to be confronted with this sarcasm! Enough said.

> *Tenth Letter*
> You've seen, said he, in everything I've told you until now, with what success our priests have laboured to discover through their enlightenment that there are a great number of things permitted that used to be forbidden; but since there are still some sins that they haven't been able to excuse, and since the sole remedy is to confess them, it has been necessary to smooth away the difficulties, in the ways I'm about to tell you. And so, after having shown you in all our previous conversations how we have soothed the scruples that used to trouble people's consciences, showing that what was thought bad is not, I have to show you the way to easily expiate what is really a sin, by making confession also as easy as it used to be difficult![15]

This letter opens with Montalte expressing his concern for the Jesuits having softened confession and thus his complaint concludes with his claim that 'the priests have spared men the *bothersome* obligation of actually loving God.'[16] What a saving of time and effort, indeed!

In particular, he takes up the doctrine of Père Pinthereau, who will let you judge the worth of this dispensation by how much it costs, which is the blood of Jesus-Christ . . . 'It was reasonable', he says, 'that in the gracious law of the New Testament God removed the bothersome and difficult obligation, that was the strict law, to accomplish an act of perfect contrition in order to be justified and that he instituted sacraments to be a supplement to this lack with an easier disposition. Otherwise for the Jews, who were slaves, to have their Lord take pity on them.'[17] Pascal, via Montalte, finds all of this troubling, especially since it seems to unfasten the words of Christ himself, who said that 'whoever doesn't love him, doesn't keep his commandments!'[18] Montalte wonders how those who never loved God during their lifetime would become worthy to enjoy him in eternity, claiming 'that's the mystery of iniquity accomplished'.[19] He bids those impressed by Jesuit casuistry to open their eyes and let these excesses cure them of their blindness.

In sum, the *Lettres provinciales* are of major importance for their style and their satirical sharpness, and have come down to us with a bevy of anecdotes surrounding them. At the time of Pascal's death in 1662, Marguerite Périer reported that in response to the query of whether he repented of having written the *Provinciales*, she heard him declare: 'I say that, far from repenting of having written them, if I had to do them over now, I would make them still stronger.'[20]

PASCAL'S FAMOUS POEM

Later on, when Pascal was somewhat reconciled to his sister's incorporation – in the literal sense – into Port-Royal, he was of course on the side of the Jansenists, and opposed to the forcing of the signing of obedience to the clergy. It is all the more interesting to have Pascal's own poem, 'Rondeau', found with the *Lettres provinciales*, about the two attitudes, with its accusation of 'laxity' on the part of the Jesuits as opposed to the severity of the Jansenists. It is full of double and triple entendres about 'Escobar' (referred to in the *Provinciales*, because we actually have some of his texts) and other intricately ironized references, and is worth quoting for its wit. Everything about Pascal's deliciously smarmy poems against the Jesuits delights us, like the ditty about 'the obliging conduct / Of this accommodating bunch of persons, / Who, to please the vicious ones / Just make the path to heaven broader' (*Telle est la conduite obligéante / De cette troupe accommodante / Qui pour complaire aux vicieux / Élargit le chemin des cieux*).[21]

Rondo:
To the Reverend Jesuit Fathers about their Accommodating Morality
Withdraw, Sins. Incomparably skilled,
The famous troop in Escobars filled
Leaves us your sweetness missing their deadly venom:
We savour them innocently, and this new way
Leads effortlessly to heaven in a deep peace.
Hell loses its rights this way, and if the devil scolds,

We just have to say to him: 'Now, filthy ghost,
In the name of Bauny, Sanchez, Castro, Gans,
Tambourin,
Withdraw.'
But, oh flattering Fathers, a fool who relies
on you,
For the unknown author rebelling in Letters
against you
Of your politics has discovered the subtle opposing,
Your Probability calculus approaches its goal.
We did all that; seek a New World for you,
Withdraw.[22]

As a counterpart to the satire against the casuistry and lax manners of the Jesuits, we might consider here the very serious relation of a miraculous cure: Pascal never took up the arms of satire against the miraculous.

THE MIRACLE OF THE HOLY THORN

Miracles might be considered a topic on which Pascal had much to say. He writes about them in his *Pensées*, distinguishing between the miracle itself and the relief that comes with it. One does not follow God because his miracles fill one with the good; rather one follows him because of his miracles and the power they evoke. In one of the *Pensées*, Pascal speaks about the thorn from the crown of Christ – the Sainte-Épine or the Holy Thorn – as a sacred relic of God's conspicuous power, considering his niece Marguerite Périer's documented cure by having touched it:

> These are not men who perform miracles by an unknown and doubtful power [*vertu*], requiring us to make a difficult decision. It is God himself. It is the instrument of the passion of his only son, who, being in many places, chooses this one, and draws people from all quarters to receive this miraculous relief for their weaknesses.[23]

About this miracle occurring on 24 March 1656, the Port-Royal nuns found it all the more marvellous for the monastery because the introit of the Mass of that day had these words: 'Make a sign in my favour, so they will see', and lo and behold, this was indeed a sign. Racine comments on the miracle, in his *Abrégé de l'histoire de Port-Royal*, speaking of Marguerite:

> She had been afflicted for three years with a lacrymal fistula in the corner of her left eye. This fistula, very big outside, had ravaged the inside. It had entirely rotted the nasal bone and pierced her palate . . . You couldn't look at her without a kind of horror . . . a priest had gathered some thorns which were supposed to have been in our Saviour's crown, and gave them to the nuns, who put one thorn inside their choir on a little altar against the grille, and a procession was made in its honour after vespers, and they went to kiss it. The mistress of the *pensionnaires* told the little girl to touch her sick eye to the thorn, which she did, declaring afterward that the Holy Thorn cured her. After this ceremony, all the other *pensionnaires* retired. No sooner was she there than she said

> to her companion: 'My sister, I'm not sick any more, the Holy Thorn has cured me.' In fact, her companion, having looked at her carefully, found her left eye as healthy as the other, without any extraneous thing and even without a scar.[24]

Even if this miraculous cure had not been visited on a family member, Pascal would say, as he does in his *Pensées*, that within all miracles, the truth prevails. He claims that since the Church's enemies are lacking in miracles, they say that 'doctrine must not be judged by miracles, but miracles by doctrine.'[25] For Pascal, miracles, the strongest proof of God's existence, are also necessary for the demonstration of faith by irrational belief.

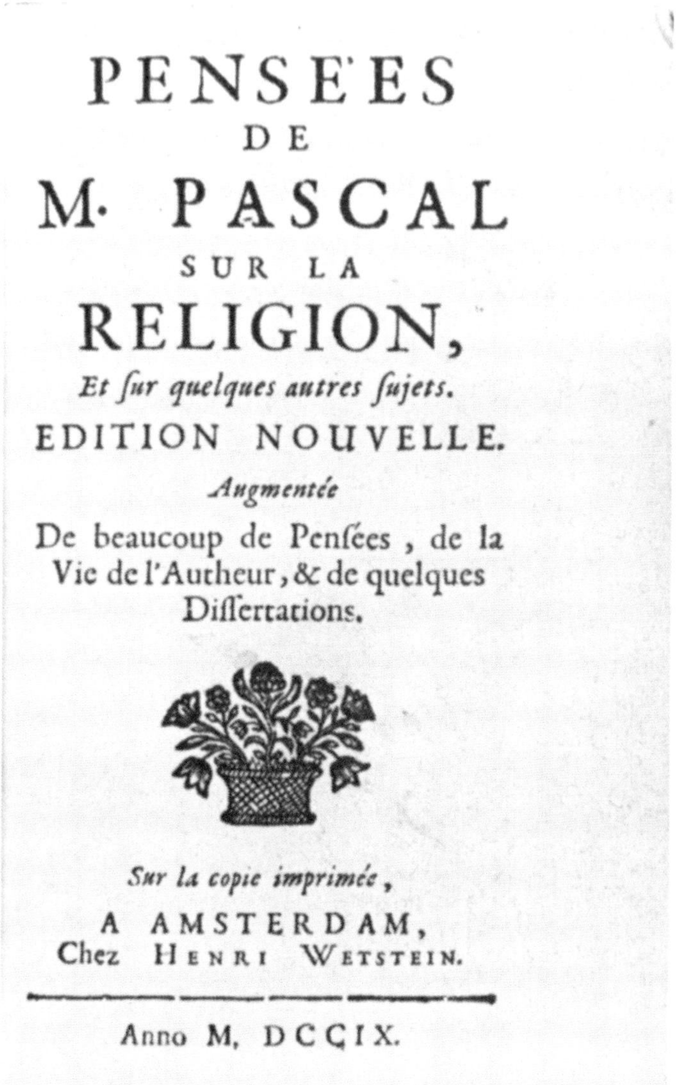
PENSÉES
DE
M. PASCAL
SUR LA
RELIGION,
Et sur quelques autres sujets.
EDITION NOUVELLE.
Augmentée
De beaucoup de Pensées, de la Vie de l'Autheur, & de quelques Dissertations.

Sur la copie imprimée,
A AMSTERDAM,
Chez HENRI WETSTEIN.

Anno M, DCCIX.

25 Title page of the *Pensées* (1709 edition).

SEVEN

Thinking about Thinking: The Publication of the *Pensées* in 1670

For Lucien Goldmann, Pascal is 'the first modern man'.[1] He states that between 1654 and 1662, Pascal moves from 'the centralist intellectualism of the *Provinciales* to the tragic extremism of the *Pensées*'.[2] Reflecting on the reflections, or *Thoughts*, that are the *Pensées sur la religion*, in these 'tragically extreme' meditations, however we read them, we see Pascal investigating various ways of approaching the problems raised by human existence.

It seems, to a present-day reader, as if we were *invited* to be thinking personally, historically, politically, persuasively, stylistically, practically and philosophically at once, with all the interruptions and fragmentary nature of our personal ruminations both excused and welcomed. This permission is given by the fragmentary form of the *Pensées* themselves. In their various orderings, they read with equal strength, new discoveries abounding. Pascal himself was of course concerned about the ordering of them towards his *Apologia*, as has been stressed before. Equally important is the fact that modern readings have tended to drop the subtitle about 'religion' from the *Pensées*, for this already says something about present-day thinking. Whatever our inclination towards the certain beliefs held by Pascal, we cannot overlook his own inflection of his thoughts and writings towards this *Apologia*, even as we read the *Pensées* in our own fashion.

ASTONISHING STUFF

These fragments manifest a variety of styles in their relation to Pascal's thoughts about the human condition. Their own condition has been of much concern to their various editors. These aphoristic fragments were originally written on large pieces of paper, divided by a stroke of the pen to mark their separation, and then cut apart in 1658 and – some of them – classified in separate batches. These latter were preparations for the reading he did for his friends in October or November that year, to show them the outline of what he was planning to bring to fruition, an *Apology for the Christian Religion*. But by late January 1659, stricken by a sickness that prevented his further work on them, he never again took up their organization. The remaining parts are 34 series of non-classified texts. The version generally used in the present book is the work of Louis Lafuma, whose preface assures us that what we are reading is the work of Pascal himself, not of the friends who then helped with future editions.[3]

As Marc Fumaroli points out in his preface to Pascal's *De l'esprit géométrique et de l'art de persuader*, the late Latin *pensare* gave two versions in modern French: *penser*, to think, and also *peser*, to weigh on a scale, so that *penser* is to put two objects of experience in balance.[4] Among the elements weighing into the *Pensées* is the thought of Pascal's great predecessor Michel de Montaigne. As Fumaroli says, 'It is hard to surpass Montaigne in pessimism about the powerful and the humble condition of the great.'[5]

Much of Pascal's writing is in response to and in argument with Montaigne, and, therefore, with himself; for instance, see *Pensée* no. 689 in the unclassified part of the texts: 'Ce n'est pas dans Montaigne mais dans moi que je trouve tout ce que j'y vois' (It is not in Montaigne, but in myself, that I find everything I see in him).[6] For a salient example, the entire text of Pascal's *Art de persuader* is a response to Montaigne's *Essais*, themselves a gathering of 'exercises' (in the sense of 'philosophical work upon the self' with the aim of undoing egoism).[7] The weighing of one consideration against another, of one kind of exercise against another, counts heavily in the presentation of Pascal's 'thoughts', and is part of the 'esprit de géométrie', quite as essential as the 'esprit de finesse' – structure, clarity and disambiguity on the one hand, and experience, intuition and non-linear thinking on the other, those contraries kept in balance throughout. The rich interpretation of the term *pensée* confers a new density upon Pascal's use of the term in his remarkable *Pensées*, in every way a stylistic tour de force.

ORDER AND FORM

Of all the legacies of Pascal the French philosopher, the *Pensées* remain the most extraordinary and the most modern. I have translated, in an informal fashion, many of the fragments selected here, in general using the Lafuma numbering, often with the Sellier numbering in parentheses following. Here Pascal reflects on the order of his thoughts or *Pensées:*

> Let no one say that I haven't said anything new – the way the matter is placed is new. (696 [22])

> Words differently arranged give a different meaning. And the meanings differently arranged give different results. (784 [23])

And yet, of course, he knows the impossibility of choosing any single one. It seems to me that this gives us permission, as do the fragmentary form and the unlikeness of the different fragments – from aphorism and incomplete reflection to lengthy meditation on the human condition – to worry a bit less than is the custom of Pascalian critics about the different orderings, those of Léon Brunschvicg, of Louis Lafuma and of Philippe Sellier. If Pascal didn't worry too much, why should we?

> Order . . . No human knowledge can hold to it. St Thomas didn't hold to it. Mathematics can hold to it, but is useless in its depth. (694 [61])

I give myself permission, then, for one further reflection on the so frequent comparison between Pascal and his great predecessor Montaigne. It isn't just that Pascal abhorred 'the crazed project of depicting himself'[8] but that the absolute contrast in their form is riveting. The easy flow of Montaigne's essays, and the fragmentary nature of Pascal's *Pensées*, like so many other contrasts between these forever enemies and associated thinkers, link them together in their genius. These extremes touch each other, as happens in Pascal and his thought: 'Les extrêmes se touchent.' How are these brilliant bursts or uneven meditations, which he jotted down on large sheets, cut up and left – some bundled together, some not – to be ordered? In an article on 'Pascal's Fragmentary Thoughts: Dis-order and its Overdetermination', Domna Stanton remarks on how their digressive and discontinuous nature is also that of the speaking subject, how 'these diverse symptoms of discursive dis-order seem designed to persuade readers of the existence of a metaphysical order that is digressive in its scriptural incarnation. Discontinuous in its own right.'[9] Their very brevity is, she points out, akin to the Stoics' *sententiae*, and their very dis-order is determined by the conviction that 'the mathematical or geometrical mind which demonstrates "par ordre" is viewed as counterproductive to the author's expressed intentions itself,' depicting the 'immanent nature of truth as contradiction, ambiguity, and uncertainty'.[10] The poetic text, as discussed in Umberto Eco's *L'Oeuvre ouverte*, is against totalization, the whole and the true order only discoverable in God. And Lucien Goldmann sums up the aesthetics and the necessity of the fragmentary form:

> Paradox is the only valid form for the expression of a philosophy that holds truth to lie in the meeting and reunion of opposites, and the fragment is the only valid form for a work whose essential message is that man is a paradoxical creature.

It is, he says, a matter of all or nothing, the only possible stance for what he calls a tragic work, 'complete by the very lack of completeness'.[11]

The great poet of modernity Stéphane Mallarmé was himself given to the constant ordering and reordering of his reflections. In the work towards which all other works were to lead, *Le Livre* (The Book), the passages were shuffled and reshuffled at every performance of its reading; thus, spontaneity and chance were essential parts of the textual adventure. It seems to me that the Pascalian venture and this symbolist/modern adventure meet on the ground of these happenings, if I can put it like that. And for many of us, the aphoristic statements in Ludwig Wittgenstein's *Zettel* (or his *Nachlasse*, his left-behind reflections that were not fitted in anywhere) play that role of liberated and liberating spurs to thought.[12] They are numbered, as the Pascal *Pensées* bear numbers, but that is not the way they sit in our mind.

BRIEF HISTORY

Pascal gave no title to these fragments that we call his *Pensées*. There was to have been a preface to them by Nicolas Filleau de La Chaise, but upon the disapproval of the family, the preface was simply printed in 1672 as a *Discours sur les Pensées de M. Pascal*. Filleau says it was based partly on a two-hour lecture by Pascal that he had heard eight years before. The discourse attributed to him takes as an example the Great Fire of London of 1666, but Pascal had, of course, died before that fiery example. Jacques Chevalier, in his Pléiade version of Pascal, had founded his own classification of the *Pensées* on the ordering of Filleau, who had read them in the first copy, and whose witness is worth listening to. Filleau reflected on those writings, remembering some of Pascal's conversation, so they are inserted in Le Guern's Pléiade edition, among the other attributions to Pascal.[13]

The title *Pensées* is of course not of Pascal's doing. Since he sewed some of them together in bundles, we might relate that action to his sewing his secret Memorial or 'testament' into his jacket's lining, transferring it and re-sewing it if he changed jackets.[14] This is faithfully reported by a servant in his house, who found the insertion in his jacket on a piece of parchment in which was enfolded another piece, the Memorial inscribed on both. For the eight years before his death Pascal unsewed and re-sewed this testament to his night of revelation into every garment he would wear. Then in 1658 – according to Philippe Sellier, generally agreed upon as the most reliable editor – he assembled a core of 27 bundles, and left seven others aside. This collection remained in the possession of his sister Gilberte and her family. Of course, as in any such matter, the question arose as to publication. The nephew of Pascal, Étienne Périer, prefaced the 1670 or 'Port-Royal edition' by alleging there was no sequence in the bundles and little coherence, to such an extent that they hesitated to publish them at all, simply copying them out as they were, in two sets, one for themselves and one for Port-Royal. As for publication, they simply took 'the ones that seemed clearest and most complete . . . [and] put then in some sort of an order and placed under the same titles those that were about the same subjects, and . . . suppressed all the others that were either too obscure or too imperfect'.[15] As the family was eager to stress the ways in which these thoughts made part of Pascal's religious *Apologia*, they simply omitted sections that did not fit in the plan or seemed to make no sense. It is on a par with their asserting that after his revelation and Memorial, Pascal no longer interested himself in the scientific world, and simply consecrated himself to the religious life. It is absolutely true that he continued to live in utmost poverty, to cherish his sickness as something sent by God, and to eschew all mortal advancement or concern. But it cannot be alleged that he deserted his scientific interest altogether.

There are three hundred years' worth of different subjective orderings, the most famous up to recent times being that of Léon Brunschvicg in 1905 and – from my own subjective point of view – the most interesting being that of Louis Lafuma in 1962, using copy A, the one made for Gilberte, and following the holes of the sewing needles, it seems, a tale I haven't forgotten. This is the ordering, slightly revised, used by Le Guern in his 1998–2000 Pléiade edition for Gallimard, in volume II. Then in 1999, Sellier based his edition on copy B, the copy sent to Port-Royal, which is a continuous text, not reordered, so more faithful to Pascal's intentions. This is the version available in the more recent publications. Each publication is careful to include a concordance of the Lafuma and the Sellier ordering, so that each fragment has two numberings. Part of the fascination in the differing readings is exactly this feeling of the separate orderings and layerings, to which a reading of the English translations adds another, as if we were indeed entitled to re-readings of Pascal's thoughts in whatever sequence might – at different moments of our life and reading – make most present sense to us. In invoking this privilege, we might call upon Pascal's own stated attitude in the last two years of his so very short life about the quarrels between the Jesuits and the Jansenists: he no longer, as

26, 27 Manuscript of the *Pensées*.

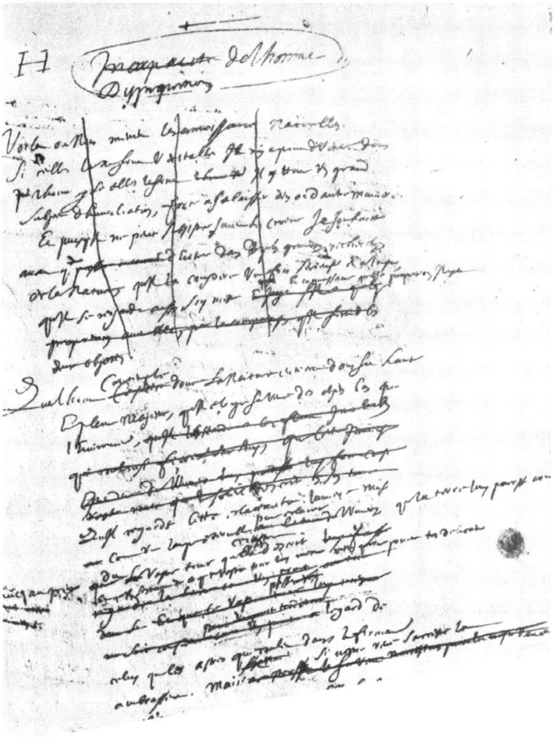

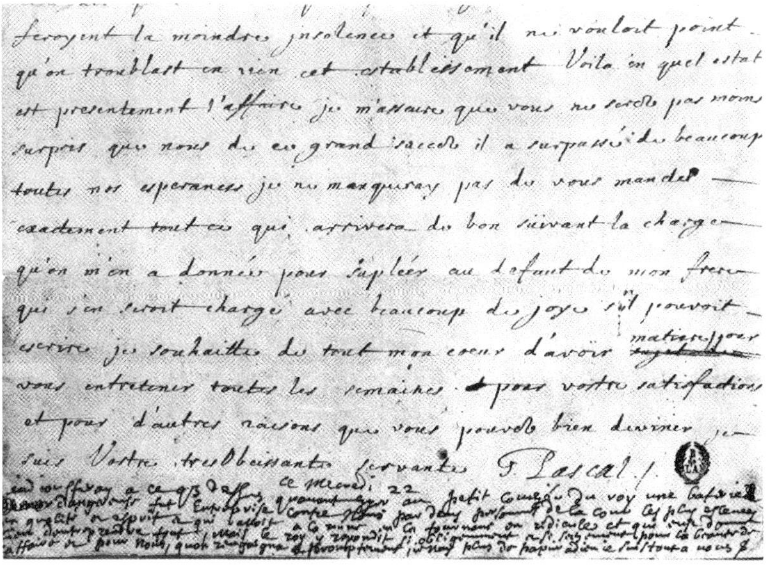

feroyent la moindres insolences et qu'il ne vouloit point
qu'on troublast en rien cet establissement Voila en quel estat
est presentement l'affaire je m'assure que vous ne serez pas moins
surpris que nous de ce grand succès il a surpassé de beaucoup
toutes nos esperances je ne manqueray pas de vous mander —
exactement tout ce qui arrivera de bon suivant la charge
qu'on m'en a donnée pour supléer au defaut de mon frere
qui s'en seroit chargé avec beaucoup de joye s'il pouvoit
escrire je souhaitte de tout mon coeur d'avoir ~~sujet de~~ matiere pour
vous entretenir toutes les semaines pour vostre satisfaction
et pour d'autres raisons que vous pouvez bien deviner je
suis Vostre tresobeissante servante G. Pascal

he states in 1661, entered those quarrels, finding them less relevant to what mattered. In the long run, that seems absolutely right. In the introduction to the Sellier version of the *Pensées*, Gérard Ferreyrolles remarks, on the term 'Jansenism', that it was a term applying to the Port-Royal group, and one that they had always rejected.[16]

In my view, it is Pascal's own inventive and provocative thoughts that most matter, however and whenever we read them. This book is based on successive readings of the available editions in their accumulative wisdom. I have used, gladly, both Pléiade editions. I am quoting from, and thinking about, the editorial complications of the two Pléiade editions, that of Jacques Chevalier, *Oeuvres complètes* (Gallimard, 1954) and of Michel Le Guern, *Oeuvres complètes*, 2 vols (Pléiade) (Gallimard, 1998–2000). I then used Sellier's version in the 'Livres de Poche' series, referring to a version published by Classiques Garnier (1991) and then the Lafuma version by Éditions du Seuil (*Essaís*) of 1962. I ended with a rereading of the Lafuma version, the first I ever read, and finally, with Roger Ariew's splendid translation of the *Pensées* (Hackett, 2004), which I have used in addition to my own renderings. It has seemed to me that the various layers of rereading might add to the density and depth of these thoughts, reordering them in my own times of rereadings and rewritings. As for the ordering, that is an enduring and fascinating question for every reader and rewriter, but I want to single out a few particular passages.

PARTICULAR *PENSÉES* AND PERSUASIVE THINKING

We could learn much, for our own styles of writing and living and thinking, from these texts, however they are arranged. Lafuma's grouping by the holes for the threading of the string exactly fits Pascal's obsessive sewing of the Memorial into his clothing, such secrecy and obsession working toward an invaluable intensity.

How to think: fragmentary bursts instead of lengthy ponderings . . .

How to talk about thinking: we could try, as Pascal does, to swivel from one thing to another, with the speed and suddenness visible not only in Pascal's writing but in his experimentation, as his biographers point out. He would move from one experiment to another, not prolonging a subject once he had seen where it was going. (See the reflections on his personality, by Albert Béguin and Jacques Attali.) He didn't need to pursue whatever it might have been initially, any further, having taken it up in its initial difficult stages. So our own thoughts might not have to exhaust themselves, but simply take themselves and us on the first and more challenging parts of the journey, once it has been sketched out. From there, we could follow, and take in our turn, a whole way of presenting a topic with sudden shifts, in a rapidity of movement, characterized by endings as sudden as beginnings. We could read some of Pascal's statements about the 'volubility' of our minds as a warning about going on too long about anything: I have tried not to go on too long about this life of this extraordinary thinker, or his thoughts.

How to weigh subjects and styles: no writer or thinker has ever been more persuasive in the use of oppositions and contraries, weighing – as Pascal does with the two modes, geometrical and intuitive – one against another, as with misery and grandeur, knowledge and ignorance. In discussing the geometrical mode, he shows how the principles are out of the ordinary habits of thinking, as opposed to the 'esprit de finesse', where the principles are right in front of our eyes, but so plenteous as to escape unless we have good eyesight. How he writes! Speaking of the 'esprit de finesse':

> 'You have to see the thing straight off, in a single glance and not through reasoning, at least up to a certain point.' (512 [670])
>
> And used to seeing and judging from a single point of view, those endowed with this kind of spirit or attitude are so astonished when someone presents them with propositions about which they don't understand anything and which, even to begin with, we have to make our way through definitions and principles so sterile that they aren't accustomed to see in such detail, and are repulsed and revolted by them.

Following on from this thought, Pascal makes a summary as to the two modes, ending again on something more inclusive, picking back up on the thought with which he began, as if he were again writing on diversion, and what is serious. The serious mode, as if it were beyond the two modes, undoes everything else:

> Geometry. Finesse. (132 [165])
> True eloquence makes fun of eloquence, true morality makes fun of morality. That is to say that the morality of judgement makes fun of the morality of the mind that is without rules. For judgement is that to which feeling belongs, as the sciences belong to the mind. Finesse is the realm of judgement, geometry is that of the mind. To make fun of philosophy is really to philosophize.[17]

How to repeat: one of the more extraordinary and undebatable facts about Pascal's presentations, thoughtful in every way and not always so obviously religious, is his art of repetition. How it is possible, we say in reading him, to find the same phrases, the same rhythms, without finding them boring? The *Pensées* do not read like sermons, but somewhat like those 'commonplace books' or compilations of proverbial sayings so many of us delight in.

To take another example, Pascal's attack on the idea of affection, as a kind of superficial attachment (as opposed, implicitly, to Christian devotion), and along with it, as an unexpected analogy but working in a perfect parallel, the comparison with other borrowed qualities, those of worldly positions and accomplishments. I have cited this before, but in a different context: 'It is unjust for anyone to be attached to me, even if it is done with pleasure and free will. I would surely betray anyone in whom I would incite that desire, for I am not the end of anyone and have nothing with which I could satisfy anyone' (fragment 15). To read this along with, just to take one example, Virginia Woolf's attack on worldly positions and responsibilities and their attendant costumings (as in Church, state and education) in her *Three Guineas* is to see how very relevant Pascal can be in relation to far later writing and thought. These *Thoughts* appeal to our own present-day thinking: that is the case I would like to make here. More relevant than those often-convoluted arguments about dates and data and orderings are his illuminating and troubling perspectives on the human mind and its surroundings, on the cosmos and on moral behaviour, on the difference between the scientific and the intuitive ways of seeing – and their interconnections, on contraries and contradictions.

TOWARDS THE *PENSÉES*

Yet none of the arguments about anything concerning him is in any sense as mesmerizing as his singular personality, his way of approaching problems and his means of expressing them and himself. Pascal's approach, the stylistics of it and the radiant energy given off by the fragmentary expression have been crucial in their influence, far beyond any religious institutional involvement. Pascal's method, his powers of digression and concentration, have remained in much human memory far longer than any debate between religious cults, between the Jansenism he espoused and the Jesuitism he despised. But all along, it is the style of his presentation that grabs our attention, and holds it, fragment after fragment, in whatever order we read them. Obsessed thoughts, like his about the human condition and the ways of perceiving our relation to it and to our enduring – and the state of grace or good works responsible for our 'salvation' – may be most persuasive in their fragmentary leavings, like bursts of thinking, not logically arranged. There are worse ways of weighing one's condition than these sparks of sparkling obsession. That is the way he started, and the way our thinking may work best. It is the style of his presentation that grabs our attention, and holds it.

To begin with Pascal exults in an art of detail. He revels, and we along with him, in the genius of the small. So it is that, in the history we read about and the life we live in, over and over, the tiny detail counts immeasurably:

> Cromwell was going to ravage all of Christianity; the royal family was lost, and his family forever powerful, except for a tiny grain of sand that got in his urethra. Rome itself was going to tremble under him. But this piece of grit being there, he died, his family went down, all in peace and the king was re-established. (750 [176])

Such details as this, these tiny grains of sand, make the translation to the greatness of the art of thinking itself – as contraries always enliven – still more powerful.

> All the greatness of humans is their thought. (759 [346])

THINKING AND DIVERSIONS

The *Pensées* have an extraordinary lightness about them, as with the wit of the great philosopher Ludwig Wittgenstein, previously mentioned; they are both serious and playful. They are deeply appealing. Take Pascal's anecdotes with their moral weights: it is quite as if you were to be telling jokes or stories with a light touch even in the midst of ultimate seriousness. The gravity of the thought and the balancing of one counter against another is, if possible, intensified by the delicacy of touch.

> Diversion – People caution everyone from childhood about caring for their honour, their well-being, their friends, and for the well-being and the honour of their friends, weighing them down with chores, with the learning of languages, with things to do, and making them believe they wouldn't be happy without their health, their honour, their fortune and those of their friends – and that a single thing lacking will make them unhappy. So they are given cares and chores that will burden them as soon as the day dawns. So you will say, that's an odd way to make them happy; what could anyone arrange better to make them unhappy? Look what you could do: you could take away all those things to worry about, because then they would see themselves, they would think about what they are, where they come from, where they are going, and they might take it more easily. And that's why, after having prepared so much for them to do, if they have some time to relax, you advise them to use it to divert themselves and play, and to busy themselves also entirely.
>
> How the heart of man is hollow and full of garbage. (139 [143])

You might not have expected this ending, with its peremptory evaluation – from diversion to hollowness, from play to garbage – it makes sense retrospectively, and in its rapid conclusion has the feeling of a leap from one place to another, from spelling out to a jabbing finality. So what we are meant to do is clearly what we are doing when we read and write such things as these; whether they lead, as in Pascal's vision, to the religious life (thus the *Apologia*) or somewhere else. But we are diverted by our own restlessness, not knowing how to stay in one place, simply. In fact, 'we never seek things, but the seeking of things' (773 [135]). And at greater length:

> Humans are clearly made to think. It is their dignity and their merit; and what they have to do is to think correctly. For the order of thought is to begin by the self, and by the self's author and its purpose.
>
> Now what do humans think about? Never about that beginning, but about dancing, playing the lute, singing, composing poetry, tilting at the ring, etc. . . . and about fighting, becoming a king, without thinking what it is to be a king and to be a human. (620 [146])
>
> What is basically wrong with being swept aside from our miserable condition by lesser things? The problem is that they are so damned unreliable, being exterior to ourselves. We may be wretched, but it is our very own wretchedness, and we aren't depending on someone or something else.

Lucien Goldmann, who makes a frequent point of the drama of our situation, tragic as it is, sees the entire life of the

Pascalian endeavour, and of Port-Royal, as if it were a theatre piece, as enacting some scene under the eyes of heaven. But it must be our own scene, in whatever situation we find ourselves, thus Pascal's saying: 'Je ne suis pas de Port-Royal.' No: Pascal, although within its walls, even as a Solitary, was chiefly himself. As we remember, in the last two years of his life, he divorced himself completely from even the quarrels between the warring factions of the Jansenists and Jesuits. He never completely separated himself from the outside world, of science or of commerce (the *Carrosses à cinq sols*) but upon nothing else, but his intimate knowingness, did he depend.

> Diversion – If someone was happy, he'd be all the happier being less diverted . . . Yes; but isn't it being happy to take pleasure in some diversion? No; for it comes from somewhere else and from outside; and so it's dependent, and always ready to be upset by a thousand accidents. (132 [170])

> We are full of things that thrust us outside ourselves. (143 [464])

A note on the same quotation reads most interestingly, to show how Port-Royal inflected Pascal's writings as they chose to:

> Montaigne *Essais* I, xix: 'The goal of our career is death, that's the necessary object of our vision: if it frightens us, how is it possible to take one step forward, without any fever? The common remedy is not to think

about it. But from what brutal stupidity can such a blindness come?'

The Port-Royal edition follows this sentence, adding a development to show how the argument fits in the apology of Pascal, 'but in God alone'(!). This addition falsifies what Pascal was doing, leading or not leading to . . . The editors simply added what they chose to, in order to inflect Pascal's thought. What editions can do and undo in their ultimate control of the text is mind-boggling. This goes right along with their calling Pascal 'the secretary of Port-Royal'!

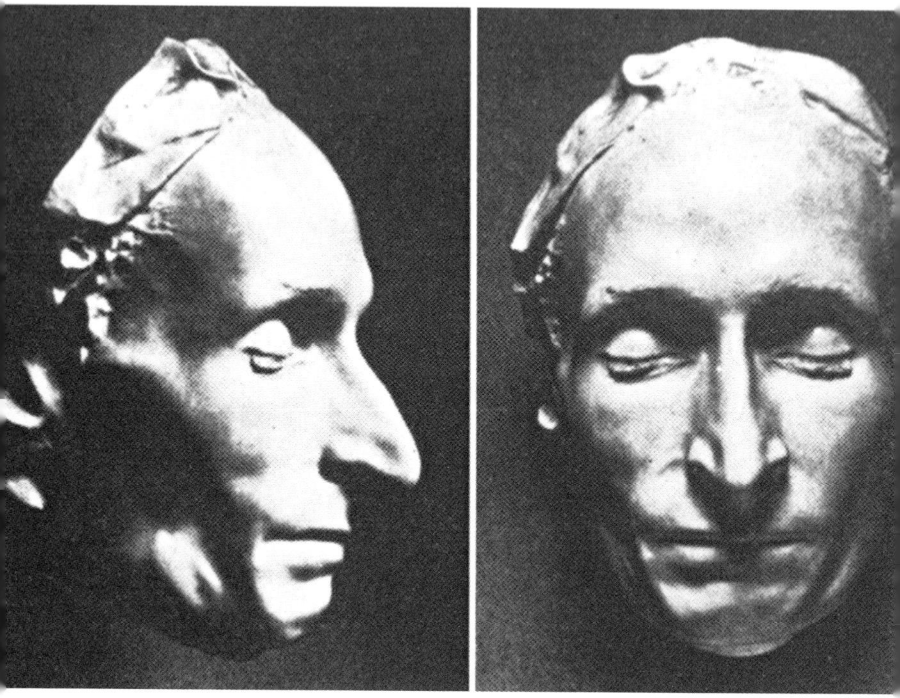

28 Death mask of Pascal.

EIGHT

Pascal's Death, and His Remaining Still with Us

As we now know, Gilberte Périer had been given the duty of recording the family happenings since Pascal, in 1662, was no longer able to write. Two years before his death, he withdrew from all religious arguments, feeling himself unable to bring any solution to the great difficulty of the questions of grace and predestination, as he confessed to Père Paul Beurier of Saint-Étienne-du-Mont. He was the Périers' parish priest and as Pascal was suffering anew, it did not matter that the curé had signed the Formulary. Pascal's retirement was bothersome to those who had wanted him to remain more strongly linked to the Jansenist cause, though he cared only to be in total submission to the vicar of Christ. His sister Jacqueline died on 4 October 1661, 'saintement' (in a saintly fashion), after the stress of the mandatory signing by the nuns of the Formulary, and on 19 August 1662, less than a year later, and after a painful time, again racked by convulsions, as Gilberte tells it, Pascal went into his death throes, uttering 'May God never abandon me' as his last words.[1] Shortly before, he had requested the viaticum, the Eucharist given to those who are dying whether or not they have been anointed, but despite his terrible headaches, they had not thought him ill enough to administer last rites. As Gilberte recounts, he said:

> Since they don't want to grant me this, I would like to accomplish in its stead some good action, and not being able to take communion in the Head, I would like to take it in my members, and so I thought of having here a poor sick person who could receive the same attention as myself. For I am discomfited by being so well helped while so many poor persons, sicker than myself, are lacking so many necessary things. Let this be done now, so that there may be no difference between him and me.[2]

When no such person could be found, he asked to be taken to the Hospital for the Incurables, to share their fate, but the doctors refused him this. He died in much agony.

On 18 November 1922, Marcel Proust, even as he lay dying, was still sticking scraps of paper into his manuscript, using his own agony to describe the death of Bergotte. He had killed himself through overwork, refusing himself all care and rest in order to write to the very end. He had given Céleste instructions about it all: 'Send for the good Abbé Mugnier half an hour after I die – you'll see how he'll pray for me.' On his last night he was dictating some reflections on death, saying to Céleste that they would 'serve for the death of Bergotte'. François Mauriac, with the other mourners after Proust's death, was 'awed by the extent of his self-sacrifice' and compared Proust's final collapse to the death agony of

Pascal 260 years earlier, as did the well-known writers Paul Morand, Anna de Noailles and Gide's friend Maurice Martin Du Gard. Mauriac, like Morand, recalled Pascal's death agonies in 1662, and especially 'the prayer in which Pascal asked God to what good use he could put illness and bodily infirmity. Marcel Proust, who was as much debilitated by pain as Pascal, like him asked the same question, and responded like him by giving his all.'[3]

Pascal was buried in the church of Saint-Étienne-du-Mont, his final parish and resting place, his body placed behind the main altar, near the Chapel of the Virgin, on the right (illus. 29).[4] The epitaph on his tomb there reads:

> Here lies Blaise Pascal, a native of Clermont, son of Étienne Pascal . . . who after having spent many years in a reclusive life and in the meditation of the law of God, died contentedly and religiously in the peace of Jesus Christ on 19 August 1662, at the age of 39. The great love he had for poverty and Christian humility would doubtless have made him wish to be without those honours that we pay to the tombs of the dead, and to live still hidden after his death, he who had always wanted to be hidden in his lifetime. But Florin Périer, his brother-in-law, counsellor of the king in the same chamber, unable to accept those desires, had this tomb built to mark the place of his burial, but more still of the piety that led him to acquit himself of this duty. However Florin is abstaining from the praise that he could bestow on him, knowing the great distancing and the aversion that Pascal had always had for such

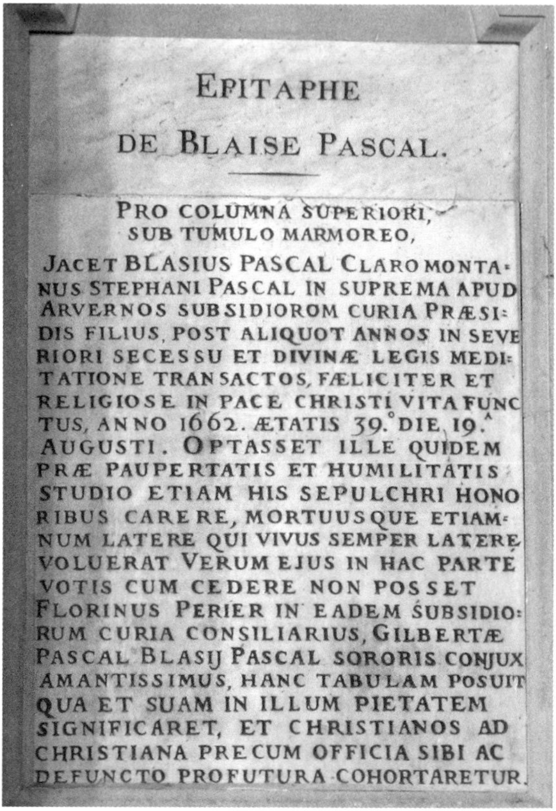

EPITAPHE
DE BLAISE PASCAL.
PRO COLUMNA SUPERIORI,
SUB TUMULO MARMOREO,
JACET BLASIUS PASCAL CLAROMONTA-
NUS STEPHANI PASCAL IN SUPREMA APUD
ARVERNOS SUBSIDIOROM CURIA PRÆSI-
DIS FILIUS, POST ALIQUOT ANNOS IN SEVE
RIORI SECESSU ET DIVINÆ LEGIS MEDI-
TATIONE TRANSACTOS, FÆLICITER ET
RELIGIOSE IN PACE CHRISTI VITA FUNC
TUS, ANNO 1662. ÆTATIS 39.° DIE 19.^
AUGUSTI. OPTASSET ILLE QUIDEM
PRÆ PAUPERTATIS ET HUMILITATIS
STUDIO ETIAM HIS SEPULCHRI HONO
RIBUS CARERE, MORTUUSQUE ETIAM-
NUM LATERE QUI VIVUS SEMPER LATERE
VOLUERAT VERUM EJUS IN HAC PARTE
VOTIS CUM CEDERE NON POSSET
FLORINUS PERIER IN EADEM SUBSIDIO-
RUM CURIA CONSILIARIUS, GILBERTÆ
PASCAL BLASIJ PASCAL SORORIS CONJUX
AMANTISSIMUS, HANC TABULAM POSUIT
QUA ET SUAM IN ILLUM PIETATEM
SIGNIFICARET, ET CHRISTIANOS AD
CHRISTIANA PRECUM OFFICIA SIBI AC
DEFUNCTO PROFUTURA COHORTARETUR.

> things, and he will simply exhort Christians to attend Pascal with their prayers, which will be as helpful to themselves as to the soul of the departed.[5]

It seems fitting that Pascal's brother-in-law, who had helped him with his experiments on the vacuum and had admired and understood the passion Pascal held for all things scientific, especially the fervour with which he devoted his life to the Jansenist cause, should erect for him such a marker.

29 Epitaph of Blaise Pascal, church of Saint-Étienne-du-Mont, Paris.

PASCAL REMAINS

> I am one of those whom Pascal upsets but doesn't convert. Pascal, the greatest of all, yesterday and today.[6]

Of all the legacies of Pascal, it is certain that the *Pensées* remain the most extraordinary and the most modern. Of particular fascination to the present generation of readers and discoverers of Pascal – easily among the most intriguing thinkers ever, a 'terrifying genius',[7] as he has repeatedly been called – is his inventiveness in his style of thinking and writing as well in all his discoveries in mathematics, science and much else. So many of the details of his life astonish and perplex as to their origin that we might tend to overlook the inventiveness of his styles, in the plural. Take, for example, his wearing a wristwatch before anyone else: other gentlemen had them in their pockets. Mlle Périer said of her uncle that he always wore a watch on his left wrist, and when she showed a watchmaker a portrait of Pascal, he recognized the portrait, and said it was a gentleman who came in often to have his watch fixed, but he didn't know his name. Pascal's being haunted by time, as by a timepiece, recurs. Take fragment 534 (457) in Ariew's translation:

> Those who judge a work without any rule stand in relation to others as those with a watch in relation to others. One says, 'It has been two hours.' – The other says, 'It has only been three-quarters of an hour.' – I

> look at my watch and say to the first, 'You are bored,' and to the other, 'Time goes fast for you, because it has been an hour and a half,' and I laugh at those who tell me that time passes slowly for me and that I am judging it according to my fancy.
>
> They do not know that I am judging it according to my watch.[8]

And those styles may be what most intimately remain with us. The example of his *Provinciales* is an obvious one, and not only for the debate between Jesuits and Jansenists, but more precisely for its style itself. What a model for oppositional politics!

Of course, his wager and the art of the game it introduces, couched in terms to appeal to the gaming instinct of Pascal's moment, is about risk-taking; so these and the discussion have implications for our present-day fascination with the stock market. Many readers are, potentially, gamblers, so discussions about risk-taking have implications for our present-day fascination with the stock market. Besides, Pascal introduced a primitive form of roulette and the roulette wheel in his search for a perpetual-motion machine. Even this invention plays into the notion of gambling, as well as its verbiage, to the excitement of gambling and its mechanisms. In retrospect, everything about Pascal is exciting – not just the wager and its relevance to the stock market and risk-taking, not just his witty refutation of the Jesuit morality he found so lax, not just his experimental methods or his inventions of so many things, from the arithmetic machine to the idea of urban transport – but all the rest, which many of us find more than intriguing. (Intriguing is probably a good word to use, given the intricacies of the interrelations between the inhabitants of and visitors to Port-Royal, both in Paris and in the country, as well as in the scientific and political circles in which the family moved.) His energy of mind allied with his corporeal frailty and constant illness combine in a legend and a reality that have left their mark on many ages and regions of the world, among which we might read such widely different representatives as these few.

CARROSSE À CINQ SOLS – THE FIVE-CENT CARRIAGE

One of the more remarkable ideas of this remarkable ideas-man is that of a possible mode of urban, low-cost transport: the five-cent carriage, a transport in common, and an idea at once practical and charitable. He and his great friend Artus Gouffier dreamed it up and financed it together, and had decided that any profit it made would go entirely to the impoverished. (Needless to say, in the long run, when its profitable day was done and it was taken over by the government, that is not where the profits would end up. Many documents were lost.) But it was a grand idea, and was the predecessor, like so many of Pascal's inventions, of what is now among us with a vengeance: public transport, with all its advantages and complications.

Pascal had put up posters announcing the novel invention and its use and advantages. These placards read: 'annonçant l'établissement des carrosses publics', and only these announcements remain. They can be found in the *Oeuvres complètes*;

Pascal might have dictated them, since it is his suggestions for these posted texts that were elaborated in letters written from March to July of 1662. In his letter of March 1662, he insists that 'to distinguish these public carriages from private ones, they will be marked with the shield of the city of Paris, and the conductors will be dressed in a blue uniform with the king's arms in front, and beneath them, those of the city, according to the power that they have.'[9] He suggested that despite this system being for public use, a passenger could book it for a private ride if he paid for all of the seats: six in total. And its public use would also be made a parliamentary rule, which means it could not be used by soldiers or pages or lackeys and livery, so as to assure complete comfort and riding ease for the townspeople. For those who did not live on the route, there would be several places where they could wait 'comfortably' and not for too long, since most people have somewhere to do business, or to take shelter before the next carriage.[10] Pascal's intention was to develop a system of direct routes so that a rider could 'go in one carriage from all the quarters of Paris to the others'.[11] In his notes on the announcements, Michel Le Guern aligns this selfless act of invention with the benevolent figure that Pascal had become:

> The carriages for five cents are, in Paris, the first organization of public transport . . . fundamentally a work of charity. At the same moment as the charity of Pascal for the souls is revealed in the project of an apology of the Christian religion of which only the fragment of the *Pensées* remain, it's the carriages for five cents that show his charity for the physical person.

> To continue such an institution, a strict organization would be necessary, founded on applied mathematics. We have never kept Pascal's calculations . . . They disappeared in 1691, together with the calculations about charity . . . The joining of an extreme precision with a concision very rare at this period is characteristic of Pascal.[12]

As Le Guern suggests, 'this work of a shareholder, of enormous use, is fundamentally a work of charity . . . [since] to move around in Paris when you don't have a carriage was difficult.'[13] In 1662, Pascal had asked his sister Gilberte to inform Simon Arnauld de Pomponne, one of the shareholders of the enterprise, of when the first line of carriages would start, as these were, in Paris, the first organization of urban public transport. Here is an excerpt of Gilberte's report to M de Pomponne, dated 21 March 1662, on the marvel of these carriages:

> It all began on Saturday at seven in the morning, but with a splash and a marvellous display . . . There were two commissioners of the Châtelet in their gowns, four guards of Monsieur, ten or twelve archers from the town and as many again of mounted horsemen . . . The townspeople were told to keep a steady rein on things, and if anyone cast the slightest insult, they would be punished. Then the coachmen put on their uniforms, blue as the colour of the king and the city, with the arms of both embroidered on the waist. Every fifteen minutes another carriage set out, each

> having a guard the whole day. At the same time, the archers of the city and the horsemen spread out along the way. It worked so well that there were many carriages full and even several women in them. The only trouble was that there were lots of people waiting in the street and then the carriages were full, so you had to go on foot. It even happened to me. They should perhaps have more than seven carriages.[14]

She goes on about the festive nature of the event, its 'Mardi Gras' feel, as a parade of townspeople set out both to watch the induction and to take a ride on the new coaches made for them. Even the workers stopped working to watch, she said, making it seem as though it were a feast day. Laughter, universal praise and joy ruled the atmosphere, though it may have been because the king himself had decreed that 'anyone who would say anything against it would be punished',[15] suggesting its importance to the city of Paris and the monarchy itself.

PASCAL'S INFLUENCE ALL OVER

Pascal was important for British philosophers, not so much because of the famous wager but because of his writing and style. John Locke, an empiricist who believed knowledge was acquired through sensory experience and not merely reason alone, had become familiar with the *Pensées* during his residence in France, and reflected on Pascal in his *Essay Concerning Human Understanding* (1690). His thoughts on identity and morality, particularly his concerns for the body as a single unit that is a part of our thinking conscious self, are urgently

30 Gérard Edelinck after François II Quesnel, *Blaise Pascal*, 1690s, engraving.

present when we consider Pascal's figure of the body, an analogy for the whole to its parts, as the body is to its members, as God is to individuals, and as the corpus is to its fragments. And Thomas Hobbes, whose moral and political thought turned on the idea of a social contract that values the state over the individual for the sake of society, met Pascal in 1631 when he spent time in Paris working as a tutor, long before he had written his controversial treatise *The Elements of Law* (1640). Pascal's insistence on the insignificance of the human in the universe and the scale of beings greatly influenced him. Pascal's father Étienne had introduced him to le Pére Mersenne. Hobbes returned to Paris as an exile in 1640 to rejoin the Mersenne circle.

Pascal's influence reached America: the first edition of the *Pensées* appeared in Amherst, Massachusetts, in 1829. Ralph Waldo Emerson, a theologist and leader of the Transcendentalist movement, kept a copy of them in his pew, to read when the sermon was dull. He found them 'stern & great, old-fashioned – theological but with sublime passages'. And President John Adams wrote to his successor Thomas Jefferson:

> I have read Paschall's Letters over again, and four Volumes of the History of the Jesuits . . . If ever any Congregation of Men could merit eternal Perdition on Earth and in Hell, according to these Historians, though like Pascal true Catholiks, it is this company of Loiola.[16]

As we would expect, Pascal's influence in France itself is the clearest and most visible trace we find. Let me quote the philosopher-historian and diplomat Édouard Morot-Sir, in his book *Raison et la grâce selon Pascal*. Pascal is, for him, immediately associated with Kierkegaard and Dostoevsky, 'whom we read with passion', and is a figure we rediscover through our favourite authors:

> We approved Paul Valéry, who didn't let himself be too frightened by the silence of the infinite spaces,[17] but we noticed that, too much like Voltaire, he had misunderstood Pascal. In Gide, we suspected a Pascal repressed; we felt a profound disquiet behind the cultivation of a self delighting in its own downcurve, its ambiguity, its ambivalence . . . Pascal, implacable and powerful scenographer of the human condition for the great novelists before the First World War, inspires after that, and perhaps more strongly still, those who lived the preludes and tumults of the Second Great War. The essays of Georges Bataille, Jean-Paul Sartre, Albert Camus – *L'Expérience intérieure*, *L'Être et le néant*, *Le Mythe de Sisphye* in 1942–3 – refer directly to Pascal, to the Pascal of the wager and of reason as the intelligence of the limits of human knowledge . . . Each thought, each action, is a wager on being, a hope arising at the extreme point of a rigorous negative thought.[18]

Morot-Sir continues with his evocation of the French, and this French philosopher so important to his country, and ours. We might have thought, he says, that after 1960 French thought abandoned Pascal in order to take up again the positivist dream of a science of man. But Claude Lévi-Strauss reveals, in *Tristes Tropiques* and in *L'Homme nu*, that his analysis of the mythopoetic structures takes place against a background of anguish and absurdity. Grace is sought in an ineffable beauty to which only music seems to give access. So structuralism hasn't escaped the great theatricalization of an absurd world, whose greatest scenographer for modern times remains Pascal.[19]

For another discussion by a French philosopher of Pascal's importance for us, we might read Pierre Bourdieu in his *Méditations pascaliennes*: 'We are automata as much as mind . . . Custom furnishes our strongest and boldest proofs; it inclines the automata, which pull the mind along without any thought.'[20] This is a social space as well as a physical one. In the space, says Bourdieu, so much is included, even 'all the paradoxes that Pascal assembled in the chapter of misery and greatness . . . Man recognizes he is miserable; he is then miserable because he is, but he is very great because he knows it.'[21] We are, as Pascal keeps saying, full of things that draw us outside ourselves, and Bourdieu takes on a Pascalian tone. This is, of course and unsurprisingly, the danger and the concomitant attraction of taking on, in his *Méditations pascaliennes*, such a powerful writer and thinker as this:

> Our instinct has us feel that we have to find our happiness outside ourselves. Our passions push us outside. Philosophers say in vain, 'Go back into yourselves, you will find your good there' . . . Objective

> probabilities never become determining except for an agent gifted with the sense of the game and the capacity to anticipate the future of the game.[22]

We hear, later in the book, exactly the same tone, as an echo of the great seventeenth-century writer:

> Finitude is the only certain thing in life, we put everything possible into forgetting it, throwing ourselves in diversion or in taking refuge in 'society'. We are foolish to take our leisure in the society of those like us: miserable like us, impotent like us, they won't help us. We shall die alone.[23]

THE SOCIAL ORDER

In *Blaise Pascal on Duplicity, Sin, and the Fall*, William Wood points out how, for Pascal, the original Fall both warps 'our capacity to love and to evaluate competing goods'[24] and encourages what he terms a duplicitous social order, as well as a similar political order, and that indeed his idea of citizens, 'socialized into accepting state rule',[25] pre-dates Althusser, Bourdieu and Foucault by about three hundred years. In this duplicitous social order, as Pascal was writing in his unpublished and incomplete *Writings on Grace*, the disorder of everyone's love of self is altered into a 'desire to deceive and be deceived'.

Samuel Beckett, the Irish playwright of *Waiting for Godot*, Pascal and Antonin Artaud, have a certain facial resemblance in their death masks. These larger than life figures form, in my mind, a kind of triptych: Pascal and Artaud, whose own meditations can be seen in dialogue with those of Pascal, as can the writings of Beckett. In a letter to Beckett, his devoted friend Robert Pinget wrote, on 10 August 1960, speaking of Beckett's wife: 'Suzanne feels she has to do a whole number on Sam's weaknesses, that I'm too fond of him, that I shouldn't get attached to this man who doesn't get attached to anyone, etc.'[26] Beckett had his own night of the soul, as he put it in the mouth of Krapp in that character's *Last Tape*: 'That memorable night in March . . . when suddenly I saw the whole thing. The vision at last', and in a letter to Georges Duthuit in 1949: 'This state, if you like, of which I can still only catch a glimpse, for a lifetime is not too long for us to get used to that darkness, except by mouthing extravagant nonsense.' Not that Beckett's vision was mystical, but, like the terrible wonderful night recounted by Stéphane Mallarmé, the mental landscapes converge. Beckett refers to a 'far whisper' haunting Molloy, and, speaking of his books and trying to give a voice to something in his head, 'That is the real thing. The rest is a game.'[27] It is, as so often, Pascal's night of vision as well as his game that is felt to haunt the far later writers and thinkers: his vocabulary, his perception.

FROM MATHS AND THE QUANTUM MOMENT TO POP AND ROCK

In 2014 in the *New York Times*, we were reminded of a great mathematician called Alexander Grothendieck, who died aged 86.[28] He was an advanced mathematician, a member of a number of learned societies, a monumental thinker who, as Pierre Deligne, a former student of Grothendieck, put it in *Le Monde*, 'had to understand things from the most general possible point of view . . . everything became so clear that proofs seemed almost trivial . . . and his ideas penetrated the unconscious of mathematicians'. He had been worried that the mathematical institute near Paris, the IHES, was receiving its funding from French Ministry of Defence . . . For indeed, 'the man who had advanced mathematics in the most profound ways did not believe that maths was the answer to everything. He taught us that life is more valuable than any equation.'[29] And along this thought line, we might look again at the relevant part of Pascal's letter to Fermat, about mathematics being the most sublime career of all – careers being, to Pascal, 'the most important thing we can choose – but only an occupation'.[30] He wouldn't, as he said, have taken two steps for mathematics. Given, as we know, the always crucial state of Pascal's own health, those two steps would have been difficult ones in any case.

'Disorder Rules the Universe', ran an article in the *New York Times* of 17 February 2015, referring to the uncertainty principle and its implications. Already in 1928 Arthur Eddington 'argued that the indeterminacy of the quantum universe opened the way for the reintroduction of the spiritual into the world, initiating a line of thinking that associates the paradoxes of the quantum world with the mysteries of religion. The American physicist Arthur H. Compton argued that the quantum world pointed to the existence of God.'[31] Ever more present, Pascal's use of pseudonyms might be compared to the contemporary movement Anonymous, which aims at eliminating the personal and its passion for privilege and notoriety. 'We are', says Anonymous, 'everyone and we are no one. We are Legion.'[32] Pascal was not no one, certainly, but he was more ubiquitous and more shape-shifting than many of his time. And ours. Pascal would be right, we might wager, to edit his own page on Facebook; all those networkings really stemmed from his calculating machine. He would have loved their calling it the Pascaline, just as he would have loved the unit of pressure called the Pascal. What mattered most was a kind of dialogue, between himself and what he believed. 'Joy, joy, tears of joy', he sewed the Memorial of that ultimate experience into his clothes so no one would see it, the intimacy and privacy of it being the real point of the emotional experience. As he so famously wrote, 'The heart has its reasons that reason does not know' – certainly it does, but he wouldn't have wanted such exposure as a publication of his Memorial would have given to the world. Only his faith – which he never lost an ounce of – motivated all those writings and those efforts at persuasion and those defences of the attitudes he believed in with all his heart; but his own 'esprit de finesse' demanded that they not be poured out openly like items of geometrical proof. Pascal was, as Goldmann states, 'the first modern man', as well as, as Antoine Adam says, the only really consistent Jansenist.[33]

And his influence is there even in hip hop and pop. In the preface to Jean-Louis Bischoff's *Pascal et la pop culture*, Jean-François Petit claims that 'Pascalian humanism works as a subversive power, capable of bringing forth an entirely other kind of horizon.'[34] Bischoff stresses the shrewd businessman side of Pascal, not just regarding his *carrosses à cinq sols*, 'which took a remarkable effort of communication and publicity', in the last year of his life, but also in that he tried to commercialize his arithmetic machine, offering a copy to Chancellor Séguier 'so that everyone would know that the highest personalities in the country wanted to possess a Pascaline'. Bischoff quotes John Lydon, aka Johnny Rotten, singer of the Sex Pistols, putting Pascal right up there with the gods of noise, because of the emotional events of rock concerts. In them, Bischoff sees a sort of tribal excitement, manifesting 'the desire of a divine sound . . . and in fact, the shamanism of Morrison, the heretical gnosis of the Sex Pistols, the black romanticism of the Goths, the parties of the house nation, the hardcore asceticism, the religious put-together of hip hop, testify to a request towards the supersensitive.'[35] These testimonies converge in a certain Pascalian excitement, apparent, and hidden.

Pascal's approach, the intense stylistics of it and the radiant energy given off by its fragmentary expression, have marked contemporary poetry, philosophy and a great deal else. I would maintain that it is something about the legend, the game, the inner workings and the outer appearance that provide a kind of poetry of genius. The artist Dorothea Rockburne has constructed an entire series around Pascal, and to his mysterious and compelling being the contemporary poet Jorie Graham bears witness in her long poem 'Le Manteau de Pascal' (illus. 31):

I have put on my great coat it is cold.

It is an outer garment.

Coarse, woolen.

31 René Magritte, *Pascal's Coat* (*Le Manteau de Pascal*), 1954, oil on canvas.

Of unknown origin.

It has a fine inner lining but it is
as an exterior that you see it – a grace.

I have a coat I am wearing. It is a fine admixture.
The woman who threw the threads in the two
        directions, headlong,
has made, skillfully, something dark-true,
as the evening calls the bird up into
the branches of the shaven hedgerows,
to twitter bodily
a makeshift coat – the boxelder cut back stringently
by the owner
that more might grow next year, and thicker,
        you know –
the birds tucked gestures on the inner branches –
and space in the heart,
not shade-giving, not
chronological . . . Oh transformer, logic, where are
you here in this fold,
my name being called-out now but back, behind,
in the upper world . . .

I have a coat I am wearing I was told to wear it.
Someone knelt down each morning to button it up.[36]

NOTE ON THE TEXT

The renderings as well as numberings of concordances differ widely, between the Krailsheimer and Ariew editions, referring to Brunschwicg, Lafuma, Ferryrolles and Sellier orderings, first and second copies, etc. None matters gravely to Pascal's thinking.

Translations are by the author except where otherwise indicated, to the best of my present recollection.

REFERENCES

KPA: Paige Ambroziak

Preface: Living with Blaise

1 Charles Augustin de Sainte-Beuve, *Causeries sur Montaigne*, ed. François Rigolot (Paris, 2003); Élie Faure, *Montaigne et ses trois premiers-nés* (Paris, 1926); Henri Lefebvre, *Pascal* (Paris, 1949).

2 'A book, even fragmentary, has a centre to which it is attracted: not fixed, but a centre that moves by way of the pressure of the book and the circumstances of its composition. A fixed centre, too, that moves, as if true to itself, by being the same and in becoming ever more central, ever more hidden, uncertain and imperious.' Foreword to Maurice Blanchot, *L'Espace littéraire* (Paris, 1955), p. 5.

3 Writes Jacques Darriulat: 'What became of the parchment of Pascal's hand? Nobody knows. It can be wondered why the recluses did not take greater care of a document that they themselves assumed bore witness to a divine vision. It has been speculated that the parchment was snipped into pieces and shared among the relics distributed to the friends of Port-Royal . . . Thus from the moment of its discovery Pascal's text is the object of entirely extraordinary treatment. In this way the parchment – the finished and elaborated work – and not the paper – the first sketch – was picked over to satisfy the fetishism of the faithful. Today it would be the inverse because we accord greater value to the initial writing than to its finished copy.' In 'Penser les *Pensées*: Le Mémorial', which features illustrations of each: www.jdarriulat.net/Auteurs/Pascal/PascalIntrod5.html. Darriulat notes further that for Pascal the Memorial was felt or touched and not seen;

that in its form it was a secret, a piece of paper in a parchment envelope sewn and concealed in Pascal's doublet.

4 Michel de Certeau, 'L'Étrange secret: Pascal', in *La Fable mystique: XVIe–XVIIe siècle*, vol. II, ed. Luce Giard (Paris, 2014), p. 310.

5 In French: 'Les manières de tourner une même chose sont infinies.' Certeau cites Pascal's *Traité des ordres numériques*, in *Oeuvres complètes*, ed. Louis Lafuma (Paris, 1963), p. 65.

6 Certeau, 'L'Étrange secret: Pascal', pp. 311–12.

7 Certeau citing Pascal's *Traité des ordres numériques*, in *Oeuvres complètes*, ed. Lafuma, p. 65.

8 Certeau, 'L'Étrange secret: Pascal', pp. 313–15. Readers may wonder why the pronoun referring to God is not in majuscule. For this reader, recalling poet Agrippa d'Aubigné's *Hécatombe à Diane* (written in the late years of the sixteenth century), the author played on the letter H to associate Hecate with Hell through reference to the figure of the gallows – common on topographical maps – represented calligrammatically as the letter H. Thus in the tradition of Baroque poetry there can be hellish connotations. For Pascal, whose appeal was made to humanity in general, it can be said that 'he' who is God is better in a form equal and common than proper.

9 Certeau, 'L'Étrange secret: Pascal', p. 326.

10 Ibid., p. 300.

11 Ibid., p. 331. Certeau grafts his words onto fragment 840 (Lafuma edition) of the *Pensées*, whose form and content resound elsewhere in the work on mystics, in which he notes that 'truth' is an error that is not yet known (*la vérité, c'est une erreur non encore reconnue*). See T. Conley, 'Legs de l'erreur: vers une cartographie politique', in *Rue Descartes: Revue du Collège international de philosophie* (1999), pp. 29–42.

12 Certeau, 'L'Étrange secret: Pascal', p. 334.

13 Readers are invited to consult the ongoing website that Dominique Descotes and his team have crafted in accord with reproductions the scraps and pieces of the papers on which the *Pensées* were written and from which figure 4 is taken: www.penseesdepascal.fr.

14 Here I follow Alfred Glauser, 'Montaigne et le roseau pensant de Pascal', *Romanic Review*, 66 (1975), pp. 263–8.

15 Michel de Montaigne, *Essais*, ed. Pierre Villey and revd Verdun-L. Saulnier (Paris, 1965), p. 452. In the Exemplaire de Bordeaux, whose pages are reproduced in digital facsimile in the Montaigne Project, readers can see how and where the author emended the text, such as in the two strike-throughs shown above. See www.lib.uchicago.edu/efts/ARTFL/projects/montaigne.

16 'Je serais tenté de le placer parmi nos grands artistes de lettres, ce graveur qui tient sa pointe comme une plume à écrire, ne serait-ce que pour faire sentir une fois de plus la fausseté des harmonies préétablies et des accords systématiques dont l'histoire des arts n'est pas encore débarrassée. À un moment où notre langue et notre style sont encore encombrés de lourdeurs, de lenteurs et de volutes, sa manière si pure et si dégagée devance les temps. Callot, graveur lorrain, est de la classe des grands écrivains français.' In *De Callot à Lautrec: perspectives de l'art français* (Paris, 1953), p. 41.

1 Early Upbringing and Arrival in Paris

1 As an overview of Blaise Pascal's life, we could rely on Jacques Attali's clarity and research to enhance a truly colourful reading. In preparation for his very readable *Blaise Pascal, ou le génie français*, he consumed enormous amounts of material about Pascal. He lists all his sources in his introduction and bibliography, which lists 530 works, and cites a whole bevy of Pascal scholars, working around Philippe Sellier, Michel Le Guern and Dominique Descotes, at Clermont-Ferrand, and mentions the hundreds of Internet sites consecrated to Pascal. Jacques Attali, *Blaise Pascal, ou le génie français* (Paris, 2002), p. 28.

2 See Olivier Reboul, *Kant et le problème du mal*, as cited in James Penney, 'The Tragic Subject: Pascal and the Mystery of the Transmission of Sin', *Paragraph*, XXIV/1 (2001), p. 31. Reboul claims that 'Adam's fault affects each of us like a hereditary evil, and . . . our freedom, corrupted at the origin, has a necessary tendency towards evil.' Olivier Reboul, *Kant et le problème du mal* (Montreal, 1971), p. 21.

3 St Justin, *Apology for Christians*, 1.29.

4 Clement of Alexandria, *Stromata*, 3.11.71.4, and Clement of Alexandria, *Paedagogus*, 2.10.95.3, at www.earlychristianwritings.com/clement.html.

5 Pope Damasus I, *Exposition of the Orthodox Faith*, Book IV, chap. 24.

6 Attali, *Blaise Pascal*, pp. 29–30.

7 Ibid., p. 30, referring to William Stewart, 'L'éducation de Racine. Le poète et ses maîtres', *Cahiers de l'AIEF* (July 1953).

8 Blaise Pascal, *Pensées*, trans. Roger Ariew (Indianapolis, IN, 2005), fragment 220.

9 Attali, *Blaise Pascal*, p. 30, referring to Jean Anglade, *Pascal l'insoumis* (Paris, 1988).

10 Jennifer Meagher, 'Food and Drink in European Painting, 1400–1800', in *Heilbrunn Timeline of Art History*, www.metmuseum.org/toah, May 2009.

11 Attali, *Blaise Pascal*, p. 51.

12 One of the more interesting ways of contemplating the writings and perspectives of this period depends on the mode of knowledge and the viewing of texts visual and verbal known as anamorphosis, of which a famous example is *The Ambassadors* by Hans Holbein, where, when you look from an angle, you see a skull. This sort of vision was made popular when Jean-François Niceron, a pupil of Marin Mersenne, published *La Perspective curieuse*, a theory of anamorphosis arrived at by applying projective geometry to the art of painting. A certain strategy of hermeneutics is based on anamorphosis, that is, to see the image of spirituality, we have to look from another than the ordinary point of view. Pascal's thinking between figure and contradiction is an example of his hermeneutics: if the Bible contradicts itself, the contradiction has to be resolved through a figurative passage, and so figures arise for Pascal at the sight of contradiction rather than from logical resolution, an idea perhaps that may have sprung from his extraordinary genius for geometrical figures. In Christian charity of course, all contradictions are supposed to be resolved. Again, 'les extrêmes se touchent': extremes are destined to meet.

13 Attali, *Blaise Pascal*, p. 52, referring to Gédéon Tallement des Réaux, *Les Historiettes* (Paris, 1962).

14 Presses Universitaires, *Blaise Pascal* (Clermont-Ferrand, 2001), p. 15.
15 Presses Universitaires, *Blaise Pascal*, p. 327.
16 Pascal, *Pensées*, fragment 464.
17 Attali, *Blaise Pascal*, p. 64, referring to Marguerite Périer, *Mémoire de la vie de Monsieur de Pascal* (Paris, 1854).
18 Daniel Fouke, 'Pascal's Physics', in *The Cambridge Companion to Pascal*, ed. Nicholas Hammond (Cambridge, 2003), p. 99.
19 Albert Béguin, *Pascal* (Paris, 1952), p. 5.
20 Ibid., p. 65.
21 Ibid., referring to Jacques Chevalier, *Pascal* (Paris, 1922).
22 Attali, *Blaise Pascal*, p. 70, referring to *Discours sur la religion et sur quelques autres sujets, restitués et publiés by Emmanuel Martineau* (Paris, 1992).
23 Attali, *Blaise Pascal*, p. 64, referring to Périer, *Mémoire de la vie de Monsieur de Pascal*, p. 76.

2 Rouen and the Pascaline; Science Trials and Conceits; the 'First Conversion'

1 The treatise, or drawing, of Schickard's calculating clock was not discovered until some time in the twentieth century when Johannes Kepler's biographer found two letters from Schickard to Kepler describing the machine in 1623–4. This discovery obviously stirred debate about Pascal's invention as possibly following it, and therefore a copy of Schickard's. See Jean Marguin, *Histoire des instruments et machines à calculer, trois siècles de mécanique pensante, 1642–1942* (Paris, 1994), p. 48. I would argue, however, that Pascal himself would say in reply to those questioning the order of invention, as he says in his *Pensées*, that 'the contest alone pleases us, not the victory'. From Blaise Pascal, *Pensées*, trans. Roger Ariew (Indianapolis, IN, 2004), fragment 637, p. 189.
2 Michael Williams, *History of Computing Technology* (Los Alamitos, CA, 1997), p. 122.
3 There were problems in the design of the calculator which were due to the French currency at that time. There were 20 sols in a livre and 12 deniers in a sol. The system remained in France until 1799 but in Britain a system with similar multiples lasted until 1971. Pascal had to solve much harder technical problems to work with

this division of the livre into 240 than he would have had if the division had been 100. Information from J. J. O'Connor and ed. F. Robertson, from the School of Mathematics and Statistics, in the University of St Andrews, Scotland (www-history.mcs.standrews.ac.uk/Biographies/Pascal.html).

4 In Blaise Pascal, *Oeuvres complètes de Pascal*, ed. Jacques Chevalier (Paris, 1954), p. 332, trans. KPA.

5 Ibid., p. 348, trans. KPA. For the small note on 'arvernus' in his pseudonym: in Latin *arvum* refers to cultivated land, and as such Pascal's use of what seems to be a cognate of the word – since 'arvernus' is not in the dictionary – may also be linked to the idea of a cultivated inventor.

6 Pascal, *Oeuvres complètes de Pascal*, ed. Chevalier, p. 349, trans. KPA.

7 Bruce Collier and James MacLachlan, *Charles Babbage and the Engines of Perfection* (Oxford, 2000).

8 Pascal, *Pensées*, trans. Ariew, p. 188.

9 Albert Béguin, *Pascal* (Paris, 1952), p. 5.

10 'Hydrostatic pressure', *The American Heritage Science Dictionary*, www.dictionary.com, 2 November 2015.

11 Ben Rogers, 'Pascal's Life and Times', in *The Cambridge Companion to Pascal*, ed. Nicholas Hammond (Cambridge, 2003), p. 8.

12 Ibid.

13 In Blaise Pascal, *Oeuvres complètes*, ed. Michel Le Guern, vol. I (Paris, 1998), p. 355.

14 Ibid., p. 358.

15 Ibid., p. 355.

16 Aristotle, *Physics*, vol. IV, trans. P. H. Wicksteed and F. N. Cornford (Cambridge, MA, 1957).

17 Pascal, *Pensées*, trans. Ariew, p. 125.

18 Desmond M. Clarke, 'Pascal's Philosophy of Science', in *The Cambridge Companion to Pascal*, ed. Hammond, p. 102.

19 In Pascal, *Oeuvres complètes*, ed. Le Guern, vol. I, pp. 430–35.

20 Béguin, *Pascal*, p. 5.

21 Jacques Attali, *Blaise Pascal, ou le génie français* (Paris, 2000), p. 119.

22 Blaise Pascal, *Discours sur la religion et sur quelques autres sujets* (Paris, 1992), trans. KPA.

23 In Pascal, *Oeuvres complètes*, ed. Le Guern, vol. I, p. 373.

24 Ibid., p. 375.
25 Ibid., p. 382.
26 Jean Khalfa, 'Pascal's Theory of Knowledge', in *The Cambridge Companion to Pascal*, ed. Hammond, p. 123.
27 Daniel C. Fouke, 'Pascal's Physics', in *The Cambridge Companion to Pascal*, ed. Hammond, p. 100.
28 Descartes' letter to Mersenne, dated 1 April 1640, is quoted in Jean Anglade's *Pascal l'insoumis* (Paris, 1988).
29 Quoted in Léon Brunschvicg, *Pascal* (Paris, 1932).
30 Attali, *Blaise Pascal*, p. 110, trans. KPA.
31 Jacqueline Pascal, in Pascal, *Oeuvres complètes*, ed. Le Guern, vol. I, p. 14, trans. KPA.
32 Ibid., p. 15, trans. KPA.
33 Ibid.
34 Attali, *Blaise Pascal*, p. 117.
35 Ibid., p. 150.
36 See the article by Henry Phillips on 'The Inheritance of Montaigne and Descartes', in *The Cambridge Companion to Pascal*, ed. Hammond, pp. 20–39. For this assertion, see p. 31.
37 In Pascal, *Oeuvres complètes*, ed. Le Guern, vol. II (Paris, 2000), p. 1087. Propos source listed in parentheses as Menjot, *Opuscules posthumes* (Amsterdam, 1997), first part, p. 115.
38 Marguerite Périer, 'Mémoire concernant M. Pascal et sa famille, par Marguerite Périer', in Blaise Pascal, *Oeuvres complètes*, ed. Le Guern, vol. I, p. 106.
39 Pascal, *Oeuvres complètes*, ed. Le Guern, vol. I, p. 100, trans. KPA.
40 Ibid.
41 Ibid., p. 102.

3 Jacqueline Pascal, Poet and Devout of Port-Royal

1 Blaise Pascal, *Oeuvres complètes*, ed. Michel Le Guern, vol. I (Paris, 1998), p. 16.
2 Jacques Attali, *Blaise Pascal, ou le génie français* (Paris, 2002), pp. 301–5.
3 Ibid.
4 Blaise Pascal, *Oeuvres complètes de Pascal*, ed. Jacques Chevalier (Paris, 1954), p. 483.

5 Pascal, *Oeuvres complètes*, ed. Michel Le Guern, vol. X (Paris, 1998), p. 162, referring to and quoting from Pascal, *Discours sur la religion et sur quelques autres sujets, restitués et publiés by Emmanuel Martineau* (Paris, 1992).
6 Ibid., *Oeuvres complètes*, vol. II, p. 306.
7 Ibid., p. 305, referring to and quoting from Pascal, *Discours sur la religion et sur quelques autres sujets*, trans. KPA.
8 Pascal, *Oeuvres complètes*, ed. Le Guern, p. 306, trans. KPA.
9 Ibid., p. 305.
10 Blaise Pascal, *Pensées*, trans. Roger Ariew (Indianapolis, IN, 2005), fragment 15.
11 Attali gives a brief, if condensed, history of Port-Royal and how Pascal and Jacqueline came to know about it in the autumn of 1647 through M Guillebert, Rouville's parish priest and confessor of Étienne's doctors M Deslandes and M La Bouteillerie. The priest told them about the extraordinary sermons of a M Singlin, who replaced Saint-Cyran at the abbey of Port-Royal, the latter established at Faubourg Saint-Jacques since 1625. See Attali, *Blaise Pascal*, pp. 124–33.
12 Attali, *Blaise Pascal*, p. 129, trans. KPA.
13 Pascal, *Oeuvres complètes*, ed. Le Guern, vol. I, p. 16.
14 The issue of her father's inheritance, her dowry for the convent, was in reality a non-issue, since becoming a nun or a monk is a 'mort civil', and material wealth is inessential.
15 See Louis Marin's works on representation and on portraiture, and on the logic of Port-Royal, all of great importance as well as subtlety. Of particular importance for this study: Marin, *On Representation*, trans. Catherine Porter (Stanford, CA, 2001).
16 Attali, *Blaise Pascal*, pp. 129–30.
17 Among the tensions of the abbey of Port-Royal, one of the most notable is known as the 'Journée du guichet', the *guichet* being the grille through which the nuns were allowed to talk to those outside. The story and the scandal began with the child Jacqueline Arnauld, who would become Mère Angélique. Struck by a sermon delivered by a passing monk, she reformed her abbey, as Sainte-Beuve tells the story in his *Port-Royal*. 'Monsieur and Madame Arnauld [her mother and father] . . . knocking at the door, *porte de clôture*: she opened the *guichet* and M Arnauld commanded her to open the

door; she begged him to enter into the little parlour to the side so she could speak to him through the grille, but he continued to strike and menace; Mme Arnauld called her daughter *une ingrate* . . . and he came into the *parloir*. She saw her father in a state of pain, and he spoke to her with tenderness of the past, of what he had done for her, of the interest with which he had always carried her in his heart . . . but this last time, he could only beg her to keep herself and not destroy herself by indiscreet austerity. These words were the greatest trial, and their tender accent was the hardest thing in the assault . . . She found herself weaker, not up to resisting; and feeling however she mustn't give in, in this struggle so overcoming, she lost consciousness and fell down in a dead faint.' (Sainte-Beuve, *Port-Royal*, vol. I (Paris, 1840), pp. 101–11, quoted in Françoise Hildesheimer, *Le Jansénsisme en France aux XVIe et XVIIe siècles* (Paris, 1991), pp. 36–7.

18 Léonard de Marandé, *Inconvénients d'état procédant du Jansénisme* (Paris, 1653).

19 Hildesheimer, *Le Jansénisme en France*, pp. 14–15.

20 To the Jansenists, the influence of Augustine seems to have been even greater than that of Thomas Aquinas. 'More than the philosopher, it is the man rich with his psychological and spiritual existence who continues to radiate.' He continues to be important in his spiritual existence even more than in his philosophical teaching. Even in politics, his notion of 'Christian order' stands out as of major importance. Now the orthodox interpretation of Augustinian thought considers that the rights of human will and the omnipotence of grace are reconciled in it. However, believers in pure and incontrovertible predestination also refer to Augustine.

21 Hildesheimer, *Le Jansénisme*, p. 38.

22 Ibid., pp. 39–40.

23 Quoted ibid., p. 50.

Sublime en ses écrits, doux et simple de Coeur,
Puisant la vérité jusqu'en son origine,
De tous ses longs combats Arnauld sortit vainqueur,
Et soutint de la foi l'antiquité divine;
De la grâce il perça les mystères obscurs,

Aux humbles pénitents traça des chemins surs,
Rappela le pêcheur au joug de l'Évangile.
Dieu fut l'unique objet de ses désirs constants:
L'Église n'eut jamais, même en ses premiers temps,
De plus zélé vengeur, ni d'enfant plus docile.

Sublime in his writings, sweet and simple in his heart,
Seeking truth to its very origin,
From all his long combats Arnauld came out the victor,
And maintained the divine antiquity of faith;
He pierced the obscure mysteries of grace,
Traced sure paths for the humble penitents,
Recalled the sinner to the binding Gospel.
God was the only object of his constant desire:
The Church never had, even in its first moments,
A more zealous avenger, nor a more docile child.

24 Hildesheimer, *Le Jansénisme*, p. 13.
25 Ibid. Of course, these questions arise continuously and Hildesheimer is able to state concisely and clearly the controversies and the opposing sides, that is, 'the controversy of St Augustine against the Pelagians, the Reformation which puts the accent on salvation by faith (without either Luther or Zwingli taking up precisely the problems of the relation between divine grace and human freedom), the prudent position of the Holy Fathers of the Council of Trent which will give birth to the opposite tendencies of the casuists and the Augustinians, the condemnation by Rome of the Jansenist expression of Augustinianism.' Hildesheimer, *Le Jansénisme*, p. 14.
26 Ibid., p. 55.
27 Ibid., pp. 13–14.
28 H. I. Marrou, *Saint Augustin et l'augustinisme* (Paris, 1955), in Hildesheimer, *Le Jansénisme*, p. 15.
29 Duchesse d'Orléans, *Correspondance*, ed. Ernest Jaeglé (Paris, 1880), in Hildesheimer, *Le Jansénisme*, p. 26.
30 Hildesheimer, *Le Jansénisme*, pp. 23–5.
31 Ibid., pp. 30–31. Hildesheimer quotes Gazier again: 'Everyone

knows this kind of war there has always existed between the University of Paris and the Jesuits. From the moment of the birth of the Compagnie de Jésus, the Sorbonne condemned their institution through censure, in which it declared, among other things, that this Society was born more for destruction than for edification.'

32 Racine quoted in Nicholas Hammond, 'Pascal's *Pensées* and the Art of Persuasion', in *The Cambridge Companion to Pascal*, ed. Nicholas Hammond (Cambridge, 2003), p. 236.
33 Ibid., p. 237.
34 Ibid.
35 Jean Racine, *Abrégé de l'histoire de Port-Royal* (Paris, 1994), p. 90, quoted in Hammond, 'Pascal's *Pensées* and the Art of Persuasion', in *The Cambridge Companion to Pascal*, ed. Hammond, p. 236.
36 Pascal, *Oeuvres complètes*, ed. Le Guern, vol. I, pp. 17–18.
37 John J. Conley SJ holds the Knott Chair in Philosophy and Theology at Loyola University Maryland in Baltimore, Maryland. See also his article 'RIP Jacqueline Pascal' in *America* (14 October 2013), Our thanks for his permission to reprint.
38 Ibid.
39 Conley, 'RIP Jacqueline Pascal'.
40 Pascal, *Oeuvres complètes*, ed. Le Guern, vol. I, p. 36, trans. KPA.
41 Ibid., p. 37, trans. KPA.
42 Ibid., p. 39, trans. KPA.
43 Conley, 'RIP Jacqueline Pascal'.
44 Pascal, *Oeuvres complètes*, ed. Le Guern, vol. I, p. 19. Notice the intimate form of address: 'témoignage d'amitié de toi . . . et te prie'. The veering back and forth between the personal intimate pronoun *tu* and the formal *vous* indicates the violent emotions in her letter – significant in every way. We see in her signature even, 'votre très humble et très obéissante soeur et servante, S.J.D. Sainte-Euphémie', the formal pronoun, like some statement of the formulaic importance of this letter. For example: 'Ce n'est que par forme que je t'ai prié de te . . . car je ne crois pas que tu aies la pensée d'y manquer. Vous êtes assuré que je vous renonce si vous le faites.'

4 Pascal's 'Worldly Period'

1 Jacques Attali, *Blaise Pascal, ou le génie français* (Paris, 2002), pp. 301–5.
2 Ibid., p. 302.
3 Ibid., p. 303, quoting from Blaise Pascal, *Discours sur la religion et sur quelques autres sujets, restitués et publiés by Emmanuel Martineau* (Paris, 1992).
4 Blaise Pascal, *Pensées*, trans. Roger Ariew (Indianapolis, IN, 2004), p. 215.
5 Attali, *Blaise Pascal*, p. 303.
6 Ibid., p. 304.
7 Sarah Bakewell, *How to live; or, A Life of Montaigne, in One Question and Twenty Attempts at an Answer* (New York, 2010), pp. 142–3. 'Bel air' definition from Hermione Lee, *Virginia Woolf* (London, 1996).
8 Pascal, *Pensées*, trans. Ariew, p. 215.
9 Blaise Pascal, *Pensées*, ed. Louis Lafuma (Paris, 1962), fragment 176, p. 418.
10 Ibid., p. 247.
11 Jon Elster, 'Pascal and Decision Theory', in *The Cambridge Companion to Pascal*, ed. Nicholas Hammond (Cambridge, 2003), pp. 53–4.
12 Elster, 'Pascal and Decision Theory', p. 55.
13 Hélène Bouchilloux, 'Pascal and the Social World', in *The Cambridge Companion to Pascal*, ed. Hammond, p. 211.
14 Ibid., p. 212.
15 Blaise Pascal, *Pensées*, ed. Philippe Sellier (Paris, 1991), p. 461.
16 Sellier 230/Lafuma 199, p. 167.
17 Pascal, *Pensées*, ed. Gérard Ferreyrolles (Paris, 1991), p. 167, fn 3, quoting Michel de Montaigne, *Essais*, II, 12, p. 520.
18 Pascal, *Pensées*, trans. Ariew, p. 212.
19 Attali, *Blaise Pascal*, p. 195, quoting from Pascal, *Discours sur la religion et sur quelques autres sujets*, trans. KPA.
20 Pascal, *Pensées*, trans. Ariew, p. 41.
21 Ibid., p. 212.
22 Blaise Pascal, *Oeuvres complètes*, ed. Michel Le Guern, vol. II (Paris, 2000), p. 43.
23 Ibid., trans. KPA.
24 Pascal, *Oeuvres complètes*, ed. Le Guern, vol. I, p. 154.
25 Ibid., trans. KPA.

26 Ibid., p. 165, trans. KPA.
27 Ibid.
28 Ibid., p. 166, trans. KPA.
29 Ibid., p. 1045, n. 2.
30 Tertullian, *De Carni Christi*, V.4. The full text reads: 'The Son of God was crucified: there is no shame, because it is shameful. And the Son of God died: it is by all means to be believed, because it is absurd. And, buried, He rose again: it is certain, because impossible.'
31 Blaise Pascal, *Oeuvres complètes*, ed. Jacques Chevalier (Paris, 1954), p. 509. The letter is number IV to Mlle de Roannez.

5 The 'Second Conversion': The Memorial; M de Sacy on Epictetus and Montaigne

1 Jacques Attali, *Blaise Pascal, ou le génie français* (Paris, 2002), pp. 212–13.
2 Blaise Pascal, *Pensées,* ed. Louis Lafuma (Paris, 1962), p. 418, fragment 176.
3 'Pascal's Memorial', translation by Elizabeth T. Knuth, ed. Olivier Joseph, revd 2 August 1999, www.users.csbsju.edu/~eknuth/pascal.html.
4 'Entretien de Pascal avec M. de Sacy sur Épictète et Montaigne', in Blaise Pascal, *Oeuvres complètes*, ed. Michel Le Guern, vol. II (Paris, 2000), pp. 82–98.
5 Ibid.
6 Ibid., p. 83.
7 Ibid.; latter trans. KPA.
8 Ibid., p. 84.
9 Ibid.
10 Ibid.
11 Ibid., trans. KPA.
12 Ibid.
13 Ibid., p. 87, trans. KPA.
14 Ibid., p. 88.
15 Ibid., p. 89.
16 Ibid., p. 97.
17 Ibid., p. 91.

18 See Henry Phillips, 'The Inheritance of Montaigne and Descartes', in *The Cambridge Companion to Pascal*, ed. Nicholas Hammond (Cambridge, 2003), pp. 20–39.
19 Sarah Bakewell, *How to Live; or, A Life of Montaigne in One Question and Twenty Attempts and An Answer* (New York, 2010), p. 142.
20 Montaigne as quoted ibid., p. 197.
21 Ibid., p. 147.

6 The *Provinciales* and the Miracle of the Holy Thorn

1 This is the topic in the discussion between the philosopher and the priest in the Criterion version of Eric Rohmer's film *Ma nuit chez Maude* (1969), in the discussion of 'charity' and Pascal's dour attitude to life (not just his wearing of a hair shirt but his absence of the sort of love called for in the Bible, says the priest). His rejection of the singular self for the multiple sense of self is of course related to his criticism of Montaigne, whose essays deal with the singular self.
2 Quoted in Jacques Attali, *Blaise Pascal, ou le génie français* (Paris, 2002), p. 424, from Tetsuya Shiokawa, *Pascal et les miracles* (Paris, 1977).
3 Blaise Pascal, *Les Provinciales; ou, Les Lettres écrites par Louis de Montalte à un provincial de ses amis et aux Révérends Pères Jésuites sur le sujet de la morale et de la politique de ces Pères*, ed. Michel Le Guern (Paris, 1987). For detailed information about these letters, see Richard Parish, 'Pascal's *Lettres provinciales*: From Flippancy to Fundamentals', in *The Cambridge Companion to Pascal*, ed. Nicholas Hammond (Cambridge, 2003), pp. 182–200.
4 Blaise Pascal, *Oeuvres complètes*, ed. Michel Le Guern, vol. I (Paris, 1987), p. 614, trans. KPA.
5 Ibid., pp. 614–15, trans. KPA.
6 Françoise Hildesheimer, *Le Jansénisme en France aux XVIe et XVIIe siècles* (Paris, 1991), p. 33.
7 Pascal, *Les Provinciales*, p. 21.
8 Pascal, *Oeuvres complètes*, ed. Le Guern, vol. I, p. 779.
9 Pascal, *Les Provinciales*, p. 346 n. 12.
10 Pascal, *Oeuvres complètes*, ed. Le Guern, vol. I, pp. 43, 49.
11 Ibid., p. 49.

12 Ibid., p. 54.
13 Ibid.
14 Ibid., pp. 101–2, from p. 346, n. 11: 'The irony seemed excessive to those making the revisions in the 1659 edition and they corrected it: "My Reverend Father, I said, how fortunate the world is to have you for a master!"'
15 Ibid., p. 156.
16 Ibid., p. 169.
17 Ibid.
18 Ibid., p. 170.
19 Ibid.
20 Pascal, *Oeuvres complètes*, ed. Le Guern, vol. II, p. 1091.
21 Pascal, *Oeuvres complètes*, ed. Le Guern, vol. I, p. 587.
22 Ibid., trans. author and Meriam Korichi:

Retirez-vous, péchés; l'adresse sans seconde
De la troupe fameuse en Escobars féconde
Nous laisse vos douceurs sans leur mortel venin:
On les goûte sans crime; et ce nouveau chemin
Mène sans peine au ciel dans une paix profonde.

Escobar, a well-known Jesuit. See saying attributed to Pascal: 'I read Escobar right through twice,' Krailsheimer, p. 331.
23 Blaise Pascal, *Pensées*, trans. Roger Ariew (Cambridge, 2004), p. 125.
24 Quoted in Hildesheimer, *Le Jansénisme*, p. 64.
25 Pascal, *Pensées*, p. 127.

7 Thinking about Thinking: The Publication of the *Pensées* in 1670

1 Lucien Goldmann, *The Hidden God: Study on the Tragic Vision in the 'Pensées' of Pascal and the Theatre of Racine*, trans. Philip Thody (London and New York, 1965), p. 171.
2 Ibid., p. 147.
3 Blaise Pascal, *Pensées*, ed. Louis Lafuma (Paris, 1962). This is the version I have used for this chapter. Other versions are the earlier one of Léon Brunschwicg, whose compilation is referred to by

numbers before each separate fragment, and a later one by Phillipe Sellier, which takes account of the philological discoveries in the 1990s.

4 Marc Fumaroli, 'Préface', in Blaise Pascal and Michel de Montaigne, *L'Art de conférer, précédé de L'Art de férer* (Paris, 2001), p. 12.

5 Ibid., p. 8.

6 Blaise Pascal, *Pensées*, trans. Roger Ariew (Indianapolis, IN, 2004), p. 172.

7 Fumaroli, 'Préface', p. 14.

8 Pascal, *Pensées*, ed. Lafuma, p. 308.

9 Domna Stanton, 'Pascal's Fragmentary Thoughts: Dis-order and its Overdetermination', *Semiotica*, special issue, 'The Classical Sign', LI/1–3 (1984), pp. 211–35.

10 Ibid., pp. 231–2.

11 Goldmann, *The Hidden God*, p. 196.

12 Joshua Wilner is reflecting on these aphorisms and on Emily Dickinson's alongside them. On Wittgenstein, see Marjorie Perloff, *Wittgenstein's Ladder* (Chicago, IL, 1996).

13 1608 in Pascal, *Oeuvres complètes*, vol. I, ed. Michel Le Guern (Paris, 1987).

14 It is irresistible, for those of us (are there many who are not?) haunted by Proust, to reflect on the last image in his great book about Françoise sewing and the way it opens into the massive building as handiwork: 'I should work beside her and in a way almost as she worked herself . . . and, pinning here and there an extra page, I should construct my book, I dare not say ambitiously like a cathedral, but quite simply like a dress.' Marcel Proust, *Remembrance of Things Past*, trans. C. K. Scott Moncrieff and Terence Kilmartin, and by Andreas Mayor (New York, 1982), vol. III, p. 1090.

15 See again the convincing translation and edition of *Pensées*, trans. Ariew, p. xiii.

16 Gérard Ferreyrolles, *Pensées: Présentation et notes, texte établi par Philippe Sellier, d'après la copie de référence de Gilberte Pascal* (Paris, 2000).

17 See also the already much quoted translation by Roger Ariew, in the introduction, of the two modes of thinking, worth citing and repeating.

8 Pascal's Death, and His Remaining Still with Us

1 *Album Pascal: iconographie réunie et commentée par Bernard Dorival* (Paris, 1978), p. 162. Lucien Goldmann speaks of Pascal's relation to the infamous document which Jacqueline had had to sign: 'In 1662 he died a radical and intransigent Jansenist refusing to countenance any signature of the formulary while at the same time professing his submission to the Church.' Lucien Goldmann, *The Hidden God: Study of the Tragic Vision in the 'Pensées' of Pascal and the Theatre of Racine*, trans. Philip Thody (London and New York, 1965), p. 170.
2 Goldmann, *The Hidden God*, pp. 164–5.
3 Richard Davenport-Hines, *Proust at the Majestic: The Last Days of the Author Whose Book Changed Paris* (New York, 2006), pp. 303, 306.
4 Goldmann, *The Hidden God*, p. 162. The tombstone replacing the first one, on which the inscription was soon rubbed out, was finally located in the southern bas-côte, because it had been transported during the 1789 Revolution to the Musée de Monuments Français and when it was restored to Saint-Étienne-du-Mont in 1817 was no longer in its original location. In its place, a modern copper plaque was put across from the one reminding us that after the destruction of Port-Royal and the profanation of its cemetery Racine's corpse was also buried there.
5 Blaise Pascal, *Oeuvres complètes*, ed. Michel Le Guern, vol. I (Paris, 1998), p. 13.
6 Goldmann, *The Hidden God*, p. 165.
7 Pierre Humbert, *Cet effrayant génie: l'oeuvre scientifique de Blaise Pascal* (Paris, 1947), p. 247.
8 Blaise Pascal, *Pensées*, trans. Roger Ariew (Indianapolis, IN, 2004), p. 134.
9 Blaise Pascal, *Oeuvres complètes*, ed. Michel Le Guern, vol. II (Paris, 2000), p. 528.
10 Ibid.
11 Ibid. A map of the pick-up places and the lines of service are available on p. 1293 of the same volume.
12 Ibid., p. 1292.
13 Ibid.
14 Pascal, *Oeuvres complètes*, vol. II, pp. 538–40.

15 Ibid.
16 John Barker, *Strange Contrarities: Pascal in England during the Age of Reason* (Montreal and London, 1975), p. 244.
17 This is referring to the celebrated fear Pascal expresses: 'Le silence éternel de ces espaces infinis m'effraie' (the eternal silence of these infinite spaces frightens me).
18 Édouard Morot-Sir, *La Raison et la grâce selon Pascal* (Paris, 1996), pp. 3–4.
19 Ibid.
20 Pierre Bourdieu, *Méditations pascaliennes* (Paris, 2003), p. 24.
21 Ibid., p. 19.
22 Ibid., p. 305.
23 Ibid., p. 343.
24 William Wood, *Blaise Pascal on Duplicity, Sin, and the Fall* (Oxford, 2013), p. 14.
25 Ibid.
26 Christopher Tayler, 'Under-the-Table-Talk', *London Review of Books*, 19 March 2015, pp. 19–23.
27 Ibid.
28 Edward Frenkel, 'The Lives of a Mathematical Visionary', *New York Times*, 25 November 2014, D3.
29 Ibid.
30 Pascal, *Oeuvres complètes*, ed. Le Guern, vol. II, p. 43.
31 Referring to Robert R. Crease and Alfred Scharff Goldhaber, *The Quantum Moment: How Planck, Bohr, Einstein, and Heisenberg Taught Us to Love Uncertainty* (New York, 2015).
32 James Gleick, 'Today's Dead End Kids', *New York Review of Books*, 18 December 2014, pp. 36–40, on the subject of Gabriella Coleman, *Hacker, Hoaxer, Whistleblower, Spy: The Many Faces of Anonymous* (New York, 2013).
323 Goldmann, *The Hidden God*, p. 209; and quoting Adam: it would be 'scarcely a paradox to maintain that there was only one really consistent Jansenist, and that was Pascal', p. 418.
34 Jean-Louis Bischoff, *Pascal et la pop culture* (Paris, 2014), p. 75.
35 Ibid., p. 200.
36 Dorothea Rockburne, *New Paintings: Pascal and Other Concerns* (New York, 1988). Peter Ruyffelaere, *Hieronymus Bosch*, trans. Ted Alkins et al. (Antwerp, 1987).

BIBLIOGRAPHY

Editions of Pascal

Barrault, Roger, ed., *Pensées: extraits* (Paris, 1965)

Chevalier, Jacques, ed., *Oeuvres complètes* (Paris, 1954)

Dorival, Bernard, ed., *Album Pascal: Iconographie réunie et commentée par Bernard Dorival* (Paris, 1978)

Ferryrolles, Gérard, ed., *Pensées: Présentation et notes, texte établi par Philippe Sellier, d'après la copie de référence de Gilberte Pascal* (Paris, 2000)

Lafuma, Louis, ed., *Pensées* (Paris, 1962)

Le Guern, Michel, ed., *Oeuvres complètes*, 2 vols (Paris, 1998–2000)

—, ed., *Les Provinciales, ou Les Lettres écrites à un provincial de ses amis et aux RR. PP. Jésuites sur le sujet de la morale et de la politique de ces Pères* (Paris, 1987)

Translations of Pascal

A Concordance to Pascal's Pensées, ed. Hugh N. Davidson and Pierre H. Dubé (Ithaca, NY, 1975)

Pascal, trans. Lilian A. Clare (London, 1930)

The Life of Mr. Paschal, with His Letters Relating to the Jesuits, 2 vols, trans. of *Les Provinciales*, by W. A. [William Andrews] (London, 1744)

Pensées, trans. Roger Ariew (Indianapolis, IN, 2004)

Pensées, trans. A. H. Krailsheimer, revd edn (London, 1995)

Pensées and Other Writings, trans. Honor Levi, ed. and intro. by Anthony Levi (New York, 1999)

Pascal's Pensées, trans. anon. (New York, 1958)
Pascal's Pensées, trans. H. F. Stewart (New York, 1965)
Pascal's Pensées; or, Thoughts on Religion, ed. and trans. Gertrude Burfurd [sic] Rawlings (Mount Vernon, NY, 1946)
Thoughts on Religion, and Other Subjects, trans. Edward Craig (New York, 1835)

Other works

Attali, Jacques, *Blaise Pascal, ou le génie français* (Paris, 2002)
Auerbach, Erich, 'The Triumph of Evil in Pascal', *The Hudson Review*, 4 (Spring 1951), pp. 78–9
Bakewell, Sarah, *How to Live; or, A Life of Montaigne in One Question and Twenty Attempts at an Answer* (New York, 2010)
Barker, John, *Strange Contrarieties: Pascal in England during the Age of Reason* (Montreal and London, 1975), p. 246
Bernstein, Peter L., *Against the Gods: The Remarkable Story of Risk* (New York, 1996)
Bischoff, Jean-Louis, *Pascal et la pop culture* (Paris, 2014)
Bonnefoy, Yves, *Goya, les peintures noires* (Bordeaux, 2006)
Bourdieu, Pierre, *Méditations pascaliennes* (Paris, 1997)
Certeau, Michel de, *La Fable mystique* (Paris, 1982)
Descotes, Dominique, *Blaise Pascal: littérature et géométrie* (Clermont-Ferrand, 2001)
Drury, John, *Music at Midnight: The Life and Poetry of George Herbert* (Chicago, IL, 2013)
Dyer, Thomas Henry, *The Supremacy of France and the Wars of Louis XIV* (San Diego, CA, 2015)
Force, Pierre, *Le Problème herméneutique chez Pascal* (Paris, 1989)
Goldmann, Lucien, *The Hidden God: Study of the Tragic Vision in the 'Pensées' of Pascal and the Theatre of Racine*, trans. Philip Thody (London and New York, 1965)
Gouhier, Henri, *Blaise Pascal: conversion et apologétique* (Paris, 1986)
Hacking, Ian, *The Emergence of Probability* (Cambridge, 1975)
Hammond, Nicholas, ed., *The Cambridge Companion to Pascal* (Cambridge, 2003)

Hildesheimer, Françoise, *Le Jansénisme en France aux XVIe et XVIIe siècles* (Paris, 1991)
Hunter, Graeme, *Pascal the Philosopher: An Introduction* (Toronto, 2013)
Kahneman, Daniel, *Thinking, Fast and Slow* (New York, 2001)
Kolakowski, L. *God Owes Us Nothing* (Chicago, IL, 1995)
Lewis, W. H., *The Splendid Century: Life in the France of Louis XIV* (Prospect Heights, IL, 1953)
Lundwall, Eric, *Les Carrosses à cinq sols* (Paris, 2009)
Marin, Louis, *La Critique du discours: sur la 'Logique de Port-Royal' et les 'Pensées' de Pascal* (Paris, 1975)
—, *La Parole mangée et autres essais théologicopolitiques*, trans. Mette Hjort (Baltimore, MD, 1988)
—, *On Representation*, trans. Catherine Porter (Stanford, CA, 2001)
Montaigne, Michel de, *The Complete Essays*, trans. M. A. Screech (Harmondsworth, 1991)
Morot-Sir, Edouard, *La Raison et la grâce selon Pascal* (Paris, 1996)
Nelson, Robert J., *Pascal: Adversary and Advocate* (Cambridge, MA, 1981)
Pater, Walter, *Miscellaneous Studies* (London, 1909)
—, *Selected Writings*, ed. Harold Bloom (New York, 1974)
Patier, Xavier, *Blaise Pascal: la nuit de l'extase* (Paris, 2014)
Perloff, Marjorie, *Wittgenstein's Ladder: Poetic Language and the Strangeness of the Ordinary* (Chicago, IL, 1999)
Rockburne, Dorothea, *New Paintings: Pascal and Other Concerns* (New York, 1988)
Stanton, Domna, 'Pascal's Fragmentary Thoughts: Dis-order and Its Overdetermination', *Semiotica*, special issue, 'The Classical Sign', LI/1–3 (1984), pp. 211–35
Thirouin, Laurent, *Le Hasard et les règles: le modèle du jeu dans la pensée de Pascal* (Paris, 1991)
Williams, Bernard, 'Rawls and Pascal's Wager', in *Moral Luck: Philosophical Papers, 1973–1980* (Cambridge, 1981), pp. 94–100
Wood, William, *Blaise Pascal on Duplicity, Sin, and the Fall* (Oxford, 2013)

PHOTO ACKNOWLEDGEMENTS

The author and the publishers wish to express their thanks to the below sources of illustrative material and/or permission to reproduce it.

Alamy: 4 (Artokoloro Quint Lox Limited); Bibliothèque Nationale de France: 3; The J. Paul Getty Museum, Los Angeles: 23; The Menil Collection, Houston, gift of Heiner and Fariha Friedrich: 31 © ADAGP, Paris and DACS, London, 2016.

Excerpt from 'Le Manteau de Pascal', from *From the New World: Poems, 1976–2014* by Jorie Graham, Copyright © 2015 by Jorie Graham, reprinted by permission of Harper Collins Publishers.

INDEX

Illustration numbers are indicated by *italics*.